PROGRESS IN

INFRARED SPECTROSCOPY

Volume 2

edited by

Herman A. Szymanski

Based on lectures from the
Sixth and Seventh Annual Infrared Spectroscopy Institute
Held at Canisius College, Buffalo, New York
1962 and 1963

SPRINGER SCIENCE+BUSINESS MEDIA, LLC
1964

Library of Congress Catalog Card Number 62-13472

© *1964 Springer Science+Business Media New York*
Originally published by Plenum Press in 1964
Softcover reprint of the hardcover 1st edition 1964

ISBN 978-1-4899-5387-2   ISBN 978-1-4899-5385-8 (eBook)
DOI 10.1007/978-1-4899-5385-8

Contents

Polarized Infrared Spectroscopy

C. L. Angell

Union Carbide Research Institute
Tarrytown, New York

INTRODUCTION

Any fundamental mode of motion that is accompanied by an oscillating electric moment leads to the absorption of electromagnetic radiation. Since the oscillating electric moment (transition moment) is a vector quantity, it can only interact with electromagnetic radiation that has an electric-field component in its direction. Therefore, if we wish to obtain information about the direction of the transition moment most conveniently, we must have it fixed in space, and must use plane polarized radiation. Since the transition moment must be fixed in space, we necessarily deal with solids in the study of polarized infrared spectra. These solids can be single crystals, oriented polycrystalline layers, or oriented polymer films.

The significance of polarized infrared spectra lies in the possibility of identifying the characteristic direction associated with a given infrared absorption. If we know the direction of the transition moment in space, it becomes easier to correlate the absorption with a mode of motion of the molecular or crystalline structure. In the study of the vibrational spectrum of larger molecules, where individual band identification is difficult, polarized infrared spectra aid in segregating the observed bands into the appropriate symmetry classes, to allow the selection of the fundamentals from a smaller number of frequencies. The same type of information can also be

1

obtained from the measurement of the depolarization of Raman-active lines and the investigation of infrared band contours in the vapor phase.

Polarized infrared spectra of crystals offer a useful method of attack, since each infrared-active band can be characterized according to its symmetry. A study of polarized infrared spectra can help us in the following ways: If the structure of the molecule and the direction of transition moments in space are known, this will help in assigning various fundamental modes of motion to infrared absorptions; or if the assignments are known, the polarized infrared spectra will yield directions of transition moments in space and information about the structure of the system under study. Sometimes these results do not agree with other structural determinations. In such cases changes in the previously proposed structure often result in good agreement with the infrared spectral interpretation.

The following subjects will be considered in this section: the means of obtaining polarized infrared radiation; the means of obtaining a suitable sample for our study; the interpretation of results and development of the necessary theoretical background; and discussion of the type of results which can be obtained by the study of polarized infrared spectra.

POLARIZERS

In the visible and ultraviolet regions transmission polarizers are quite easy to obtain. However, there are no such transmission polarizers available for the general infrared region. Therefore, to obtain polarized infrared radiation, we have to resort to the principle of polarization by reflection.

For radiation with the electric vector vibrating perpendicular to the plane of incidence, the intensity of radiation reflected from a dielectric is given by the formula

$$(I_\sigma)_r = I_\sigma \frac{\sin^2 (i - r)}{\sin^2 (i + r)}$$

and for radiation with the electric vector vibrating parallel to the plane of incidence, the intensity of radiation reflected is given by

$$(I_\pi)_r = I_\pi \frac{\tan^2 (i - r)}{\tan^2 (i + r)}$$

where I_σ and I_π are the intensities of the two polarized beams before reflection, i is the angle between the direction of the incident light and the normal to the interface, and r is the angle between the direction of the refracted light in the medium and the normal.* The dielectric constants of the two media determine an angle, tan i = n'/n, at which the reflection of the parallel electric vector is 0. At the same angle, called Brewster's angle, there is considerable reflection of the perpendicular electric vector. In Fig. 1 an illustration of reflection in the case of an air-selenium interface is given [26]. It can be seen that at Brewster's angle (which is about 68° in this case) the reflected radiation is entirely perpendicular. Thus we have obtained pure polarized radiation by reflection. However, using a polarizer made of a reflecting surface is rather inconvenient, since we require one that can be easily inserted into and removed from a spectrometer, without upsetting the optical path. If we now consider the transmitted radiation at Brewster's angle, all of the parallel radiation goes through while the major portion of the perpendicular radiation is lost. The perpendicular radiation can be decreased by using several reflections at the same angle, until the transmitted beam consists of parallel radiation only.

The most commonly used polarizer is made up of six silver chloride plates which are held in a frame at the required angle to the incoming beam [46]. This polarizer can be used for radiation of approximately $2-20\mu$. Another commonly used polarizer is made up of selenium films [26]. These selenium layers are only 4μ thick, and because of this extreme thinness are very fragile.

N. J. Harrick [33] has developed an infrared polarizer depending on reflections at a germanium—mercury interface with which two reflections are sufficient. This polarizer can be used with wavelengths ranging from 2μ all the way out into the far infrared. Such a polarizer is described in Harrick's article, but as far as I now know, it is not available commercially.

An infrared polarizer made of pyrolytic graphite, which could be called a true transmission polarizer, has also been described [53]. The optical anisotropy of pyrographite, resulting from its electrical anisotropy, causes light falling on this polarizer to become

---

*According to common usage the plane of polarization is defined as the plane containing the *magnetic* vector; however, in the present article we are concerned with the direction of the *electric* vector, so that throughout this discussion the direction of the latter will be pointed out.

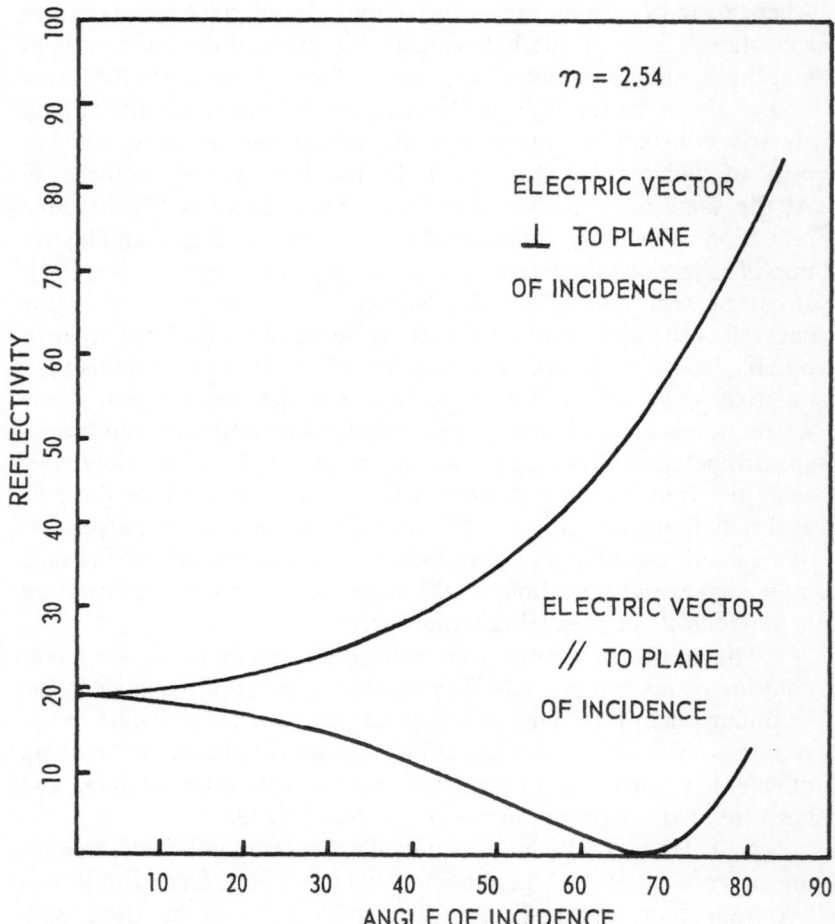

Fig. 1. Reflection at an air–selenium interface.
[Reprinted by permission from *J. Opt. Soc. Am.* 38:213 (1948).]

polarized, since the absorption coefficient has a maximum if the electric field is in the c direction of pyrographite, while it has a minimum perpendicular to the c direction. This polarizer made of pyrolytic graphite can be used with wavelengths from 10μ out to the far infrared region.

Bird and Shurcliff [13] discussed an improvement in the design of pile-of-plate polarizers, describing one made of silver chloride plates. They found that if the silver chloride plates are of uniform thickness and are exactly parallel, a number of undesired reflected rays can get through the plates, and thereby decrease the degree of polarization attained. Thus they describe a scheme (illustrated in Fig. 2) whereby the silver chloride plates used are thinned down at one end (wedging) and lined up not completely parallel to each other (fanning). The actual wedging and fanning angles required for improved performance are very small, and rolled plates of silver chloride intended to be parallel-faced will usually yield smaller sections that have the required thickness difference, so that they can be used for making up silver chloride polarizers.

SAMPLE PREPARATION

The preparation of a suitable sample is probably the most difficult part in the practice of polarized infrared spectroscopy. The study of the infrared spectrum of any material in the condensed phase requires a very thin $(4-30\mu)$ layer. At the same time, to cover the spectrometer slit a sample as large as $\frac{1}{2}$ in. $\times$ 1 in. might be needed. The preparation of a single crystal of these dimensions would be just about impossible. However, methods have been used in which single crystals prepared by the usual crystallographic means have either been cleaved or polished down to the required dimensions. Polishing is used in most of these cases since the thickness of the sample can then be reasonably well controlled.

In order to realize the possibilities of using single crystals for polarized infrared spectra, it is extremely important to keep in mind the limitations imposed on this method by the complicated manner

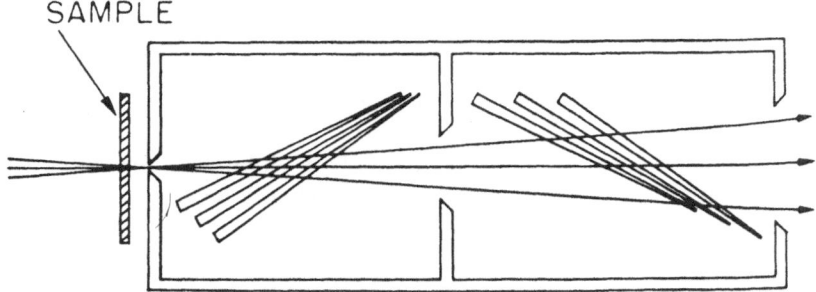

SAMPLE

Fig. 2. Schematic diagram of six-plate polarizer employing fanning and wedging. [Reprinted by permission from *J. Opt. Soc. Am.* 49:236 (1959).]

in which light is propagated through a crystal. The structurally significant directions are defined inside the medium, whereas the plane of polarization and the direction of propagation of the radiation are identified only outside the medium. If we are to use the latter directions to obtain knowledge about the former ones, we must select orientations for the crystal such that plane polarized radiation may traverse it without suffering either refraction or change of polarization character. Otherwise the incident radiation will suffer refraction according to no simple law, and the electric vector will be resolved in the medium depending on the direction of propagation and the frequency of the light.

For many crystals, the directions of the polarization axes are fixed by symmetry and become independent of the frequency of light. In uniaxial crystals of the hexagonal, trigonal, and tetragonal systems, and for biaxial crystals of the orthorhombic system, all the distinguishable polarization axes are fixed by symmetry along the crystallographic axes. For monoclinic crystals, one axis is fixed by symmetry while the other two are unrestricted. There are no restrictions upon the axes of a triclinic crystal.

Therefore, one should always remember that spectra observed along or perpendicular to polarization axes fixed by symmetry will assume a fundamental significance, while spectra taken in other directions can lead to erroneous interpretations, unless the light-transmitting properties of the crystal under study are fully taken into account.

If a single crystal is not available, then a polycrystalline mass in which the microcrystals are all uniformly oriented in one direction can be used. Such layers can be grown either from a solution by evaporation of the solvent, or from a melt by controlled cooling or by sublimation. Such polycrystalline oriented layers have the disadvantage that the layer thickness is not uniform and the orientation not always exact, but in many cases they have given satisfactory results.

Large molecules, such as long-chain polymers, proteins, or other natural long-chain molecules, can usually be oriented in thin films by depositing them from solution in certain ways, by cold or hot drawing, or by some suitable treatment. These layers usually also suffer from a lack of uniformity in thickness and orientation. In cases of naturally occurring polymers, especially proteins, samples often can be found that are well oriented in one direction.

A quantitative treatment of such partially oriented polymers will be given in a later section.

Instead of giving generalized instructions for the preparation of samples for the study of infrared dichroism, we shall give the method of sample preparation separately for each example discussed.

MEASUREMENT

The quantity we endeavor to measure in the study of polarized infrared spectra is the dichroic ratio, i.e., the ratio of band intensity when the electric vector in the radiation falling on the oriented sample is in such a direction as to give maximum intensity to band intensity when the electric vector is perpendicular to this direction.

Ambrose, Elliott, and Temple [2] give a quantitative treatment of the method for locating the direction of a transition moment in space by the measurement of polarized infrared spectra along the three major axes of the crystal. They point out that, in general, a unique direction of transition moment could only occur in crystals with one molecule per unit cell. When there is more than one molecule in the unit cell, calculations of transition-moment direction will lead to two values of either or both of the angles defining the transition moment in the crystal. In the case of acetanilide [1] four directions of the N—H transition moment were found; two of them could be eliminated as incompatible with the x-ray data, but the ambiguity concerning the other two could be removed only by comparison with other related molecules.

In order to obtain all the experimental data needed, two samples are necessary, one containing two of the crystal axes in the plane of the sample, and the other containing the third axis. If these samples are not obtainable then one crystal can be used, for example, in the following way: If the *a* and *b* axes are included in the plane of the sample, we can obtain polarized spectra along the *a* and *b* axes; if the sample is then tilted around the *b* axis, with the light falling on the sample perpendicular to the *b* axis, we can observe bands arising from spectra along the third axis. Usually the angle of tilt cannot be very large, but tilts up to $15-20°$ have been used in various experiments.

A number of experimental errors can occur in the measurement of dichroic ratios. In addition to imperfect orientation of samples, we would like to mention imperfections of the polarizer and the polarization of the spectrometer itself. This last error originates mainly from reflections at the prism face, and gives rise to the fact that the infrared beam going through the spectrometer is already partially polarized. An article by Charney [20] quantitatively discusses these two errors, and suggests methods of evaluation of possible errors in the dichroic ratio measurements due to these factors.

We have already mentioned the difficulty of preparing samples large enough for polarized infrared studies. When only very small samples can be obtained, a reflecting microscope can be used. A number of such reflecting microscopes have been described, and Fig. 3 is a general diagram of one. In this case the useful area at the illuminated spot at the microscope stage is about 1 mm × 0.13 mm. Many satisfactory spectra reported in the literature have been obtained on the reflecting microscope with both polarized and unpolarized infrared radiation. Nevertheless, a number of difficulties do arise. One of them is the strong heating effect of the concentrated radiation at the microscope stage. Cole and Jones [21] have presented a table to show that when the sample is supported on a sodium chloride plate, a compound with melting point of approximately 40°C is stable, but a compound melting around 30°C will melt. The other difficulty is due to the large convergence angle of the beam at the microscope stage. The effect of convergence on polarized infrared spectra was treated in an article by Wood and Mitra [62]. They have shown that when a uniaxial crystal has absorbing dipoles perpendicular to the plane of the sample, these dipoles can absorb convergent radiation that is falling on the sample along the unique axis. No such absorption occurs with parallel radiation. In the case of calcite the out-of-plane deformation vibration of the carbonate ion at 881 cm^{-1} can show as much as 50% absorption depending on the degree of convergence and sample thickness.

THEORY

The method described by Ambrose et al. [2] and Pimentel and his co-workers [51, 52] at about the same time is known as the oriented gas model. It assumes that the behavior of the sample is that

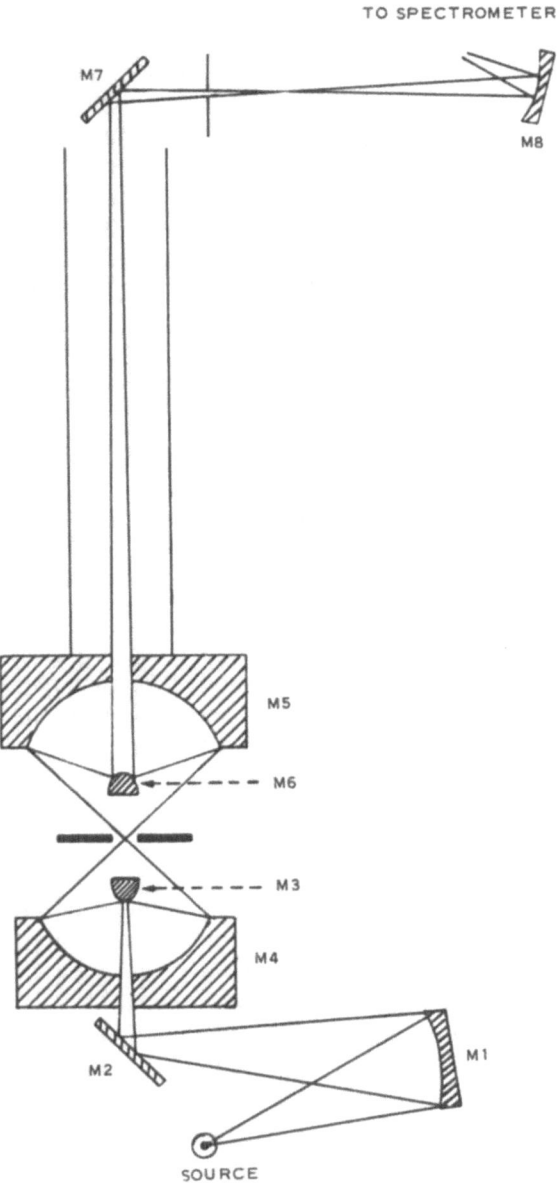

Fig. 3. Optical system of a microilluminator. [Reprinted by permission from Chemical Applications of Spectroscopy, Interscience Publishers. Inc., New York, 1956, p. 314.]

of a gas whose molecules remain perfectly oriented in space but do not interact at all. This assumption is only approximate since there are interactions in the crystal. These give rise to two different effects: the splitting of bands in consequence of the interaction between molecules having the same structure; and the violation of the gas selection rules, inactive vibrations becoming active again because of interaction of neighboring molecules. The magnitude of this interaction can usually be evaluated from the size of the shift of bands from the gas phase to the crystalline spectrum, and the size of the splitting of bands. In the case of naphthalene, for example, there are very small shifts from the gas phase spectrum to the crystalline spectrum and the splittings, where found, are very small ($2-3$ cm^{-1} only). In this case, therefore, we can assume that the interaction between neighboring molecules is very small, and as a first-order approximation we can treat the molecules as oriented gas molecules.

More rigorous treatments of the selection rules for molecular crystals have been developed by Halford [32], Hornig [35], and Winston and Halford [61]. Since recent reviews on this subject are available [44, 59], in this paper we shall present only an outline of the general principles involved.

Of the two procedures for determining the selection rules for vibrational spectra in crystals, the first (factor group analysis) proposes that all the spectroscopic frequencies can be discovered and classified by examination of the isolated crystallographic unit cell. The second method (site group analysis) assumes that one can deal with the motions of one molecule moving in a potential field that has the symmetry of the surrounding crystal. This assumption is strongly supported by experimental findings, the coupling between molecular motions in different molecules in a crystal being extremely weak in most cases.

A suitable knowledge of the crystal structure is necessary for both procedures. This usually consists of the *space group* designation along with the number of molecules per unit cell. The space group (S) is characteristic of the total crystal structure, which can be constructed from the unit cell by the translations which carry any unit cell into any other. The *unit cell group* or *factor group* (U) is characteristic of the internal crystal structure, i.e., the symmetry properties of the unit cell, if we define the translations which carry a point in a unit cell into the equivalent point in another cell as identity. The two are related by

$$(S) = (U) \times (T)$$

where (T) is the group consisting of pure translations.

About any point in the crystal there is a local symmetry. For most points this will consist only of the identity operation. However, if the point is located on some elements of symmetry, the corresponding operations leave that point invariant. Such a point is called a site. Usually a unit cell has several different kinds of sites, and sometimes it has several definite sets of the same kind. The symmetry of a site is described by the *site group,* which is a subgroup of the factor group.

The center of mass of a molecule does not change under the operations of the molecular group. Its equilibrium position in the crystal is usually situated on a site. This is only possible if the site group is a subgroup of the molecular group. The site group, in general, will be of lower order than the molecular group, although this is not necessarily so. When it is true, the selection rules for the crystal will be less strict than for the isolated molecule and, therefore, the spectrum of the crystal will be richer in bands than that of the gas phase.

When the site group has been identified, the selection rules can be deduced in the usual way. The relation between the vibrations of an isolated molecule and one in the crystal can be easily established by comparing character tables for the molecular group and the site group.

To identify the site group, one has to remember that an acceptable site group must be a subgroup of both the factor group and the molecular group. Sometimes only one common subgroup exists, but if there are more than one, a further distinction can be made on the basis of the number of molecules in the unit cell. In order to employ this procedure, it is important to have the table given at the end of Halford's article at hand. It lists the site groups for each of the 230 space groups, the numbers of equivalent sites per set, and the multiplicities of sets.

We can paraphrase Halford's summary [14] of these matters as follows: the selection rules that pertain to an *isolated* molecule are determined by the symmetry of the molecule itself, described by the *molecular group*. The selection rules that govern the molecule *in the crystal* are determined by the *static* symmetry of the field caused by its surroundings, described by the *site group*. The most complete selection rules, which involve the *dynamic* interactions of the molecule with its neighbors, are determined by the symmetry of the complete unit cell, described by the *factor* group.

At this stage we would like to present a few examples which illustrate a number of points discussed in the preceding sections.

In the case of urea, the question arose as to whether the molecule is completely planar, i.e., whether the hydrogen atoms are in the same plane as the heavy atoms. Ordinary infrared and Raman spectra did not give sufficient information to resolve this problem. However, work by Waldron and Badger [60] on a single microcrystal of urea in the reflecting microscope has given conclusive evidence for the planarity of urea.

Urea would have C_{2v} symmetry regardless whether the hydrogen atoms are all planar or above and below the plane of the heavy atoms. However, the selection rules are different in the two cases. In the first case, one would expect four N—H stretching vibrations, while in the second case only three of these would be infrared active. Waldron and Badger located four of these bands, two each in the respective polarized spectra, and thereby have shown that there are indeed four infrared-active N—H vibrations.

Polarized spectra were of considerable help in the completion of the total vibrational assignment of urea. The crystal structure of urea [17] is tetragonal and belongs to the space group V_d^3; there are two molecules in the unit cell, with the C—O bonds pointing in opposite directions and the planes of the molecules at right angles to each other.

Angell [3] has prepared an oriented microcrystalline mass of urea by melting a small amount of urea on a sodium chloride plate and then cooling it at one end. The crystals started growing at that end and were completely oriented in one direction, as could be observed under a polarizing microscope.

Figure 4 gives the polarized infrared spectrum of this oriented microcrystalline layer, along with the spectrum in a KBr pellet and the spectrum of the oriented layer using nonpolarized radiation. It can be seen that the bands at 1680, 1600, 1160, 1010, and 546 cm^{-1} are polarized parallel and the bands at 1625, 1470, 788, 714, and 600 cm^{-1} are polarized perpendicular, while the broad band

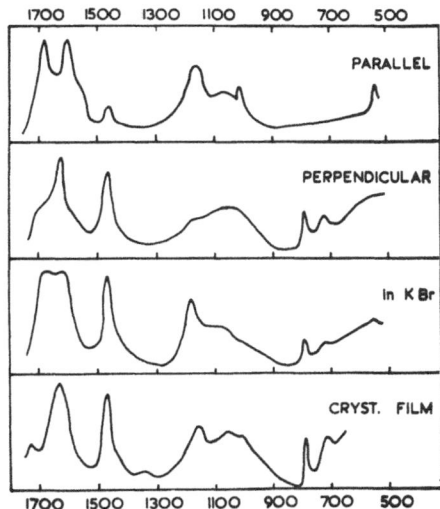

Fig. 4. Infrared spectra of crystalline urea.

extending from 1150 to 900 cm^{-1} seems to have the same intensity in both polarizations. In the polarized spectra, bands at 1625 and 1600 cm^{-1}, which previously could not be separately identified, were resolved. In this way the polarized spectra served as additional information for the assignment of fundamental vibrations. The last spectrum is of interest because it shows the effect of the polarization of the spectrometer. This spectrum taken with nonpolarized radiation shows great resemblance to the spectrum described as perpendicular, showing that the radiation going through the spectrometer had at least a partially perpendicular character.

The oriented gas model has been used by Pimentel and McClellan [51] in their interpretation of the naphthalene spectrum. A single crystal of naphthalene was grown and then polished to the required thickness. Naphthalene crystallizes in space group C_{2h}^5, and the principal cleavage plane is the ab plane. Spectra of the crystals were recorded at normal incidence for the b axis both perpendicular and parallel to the plane of polarization. Other spectra were taken at $\pm 15°$ from normal incidence, for the perpendicular orientation of the b axis.

The molecular symmetry of naphthalene is D_{2h}, while the site symmetry is C_i and the factor symmetry is C_{2h}. The factor group analysis predicts that each vibration of gaseous naphthalene will be split into two components in the crystal. After a comparison of the vapor-phase, liquid, and crystalline spectra of naphthalene it was concluded that the perturbation of the vibrational frequencies of gaseous naphthalene upon solidification is about 3 cm^{-1}, and that doublet splitting is quite small, only about 2 cm^{-1}. Since the crystal field perturbation seems to be quite small, it was possible to use the oriented gas theory.

The dichroism of nine strong bands was studied in detail by measuring the intensity of each band as the crystal was rotated, and three distinct patterns could be found. These were compared with theoretically calculated curves (dotted lines in Fig. 5), and the bands could be correlated with the different symmetry classes. The same bands studied in the gas phase could be classified according to symmetry classes by band contours. The agreement between these two methods was very satisfactory.

In a later article Pimentel and his co-workers [52] critically examined the validity of the oriented gas model in the light of their results on naphthalene. They established five spectral characteristics which can be used to test for oriented gas behavior. For naphthalene they find: "Occasional, distinct deviations from oriented gas behavior are observed, indicating that even for a molecular crystal involving relatively weak intermolecular forces the model is not completely applicable. On the other hand, it is evident that the dichroic properties of solid naphthalene are most often consistent with the predictions of the model."

Another recent application of polarized infrared spectra was the vibrational assignment of the infrared-active modes of anthracene and anthracene d-10 [18, 19]. Single crystals of anthracene were prepared by sublimation; thin sheets 25–100 mm^2 in area developing along the ab plane were easily obtained. The crystals were then examined with the electric vector of incident light parallel to the a or to the b crystal axis.

Anthracene crystallizes in the monoclinic system C_{2h}^5 with two molecules in the unit cell. The molecular symmetry is D_{2h} and the site group symmetry is C_i, as for naphthalene. According to the selection rules for an isolated molecule, the dipole moment variation takes place along the long molecular axis for B_{3u} vibrations, along the short molecular axis for B_{2u} vibrations, and along the normal to the molecular plane for B_{1u} vibrations. Vibrations of the B_{3u} species should be strong with the light polarized perpendicular to the ab

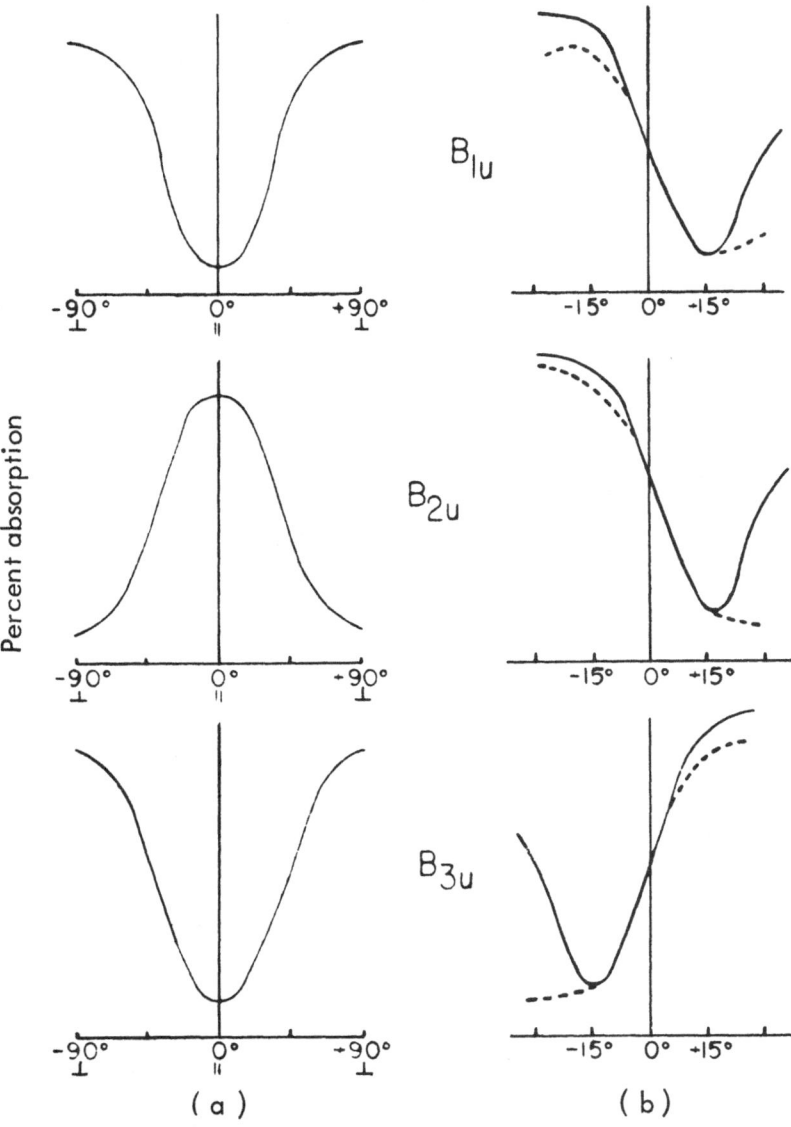

Fig. 5. Dichroism of the cleavage plane of solid naphthalene. (a) Effect of rotation of the *b* axis with respect to the plane of polarization; (b) effect of change of the angle of incidence. ——— experimental, -------- calculated dichroism. [Reprinted by permission from *J. Chem. Phys.* 20:277 (1952).]

plane, and extremely weak with the light polarized in the *ab* plane. If, as in the present case, the *ab* plane of an anthracene crystal is illuminated with polarized light, B_{3u} bands should be barely detectable. On the other hand, they should become evident in the spectrum of a disoriented specimen, for instance, in a KBr pellet. Essentially the assignment of the B_{3u} modes is based on this argument. When one looks at the table in which Califano summarizes the assignment of the infrared-active frequencies of anthracene, one realizes the importance of the polarized spectra in achieving this interpretation.

We would now like to give two examples of site group effects. The first of these is the study of the infrared spectra of single crystals of ammonium nitrate by Newman and Halford [47]. In this study, they discuss several questions: the general theory of crystalline spectra; the verification of assignments of frequency to the molecular modes of vibration; the degree of disorder resulting from rotation or other causes; and the verification of the complete crystal structure for one modification of NH_4NO_3.

Single crystals of ammonium nitrate (IV) were obtained by slow evaporation of an aqueous solution at 25°C in the form of orthorhombic crystals. Such single crystals were then polished so as to provide samples with the principal faces parallel to the (100) and (010) planes. The crystal structure of NH_4NO_3 (IV) is of the space group V_h^{13} and has two molecules per unit cell. The site group is C_{2v} for both the nitrate and the ammonium ions (see Fig. 6) and the molecular symmetry of the nitrate ion is D_{3h}.

The effect of the site group is shown, for example, in the case of the fully symmetric breathing vibration of the nitrate ion, which is inactive in the infrared in D_{3h} symmetry, but becomes active in C_{2v} symmetry and is polarized along the *a* axis. This is actually observed in the spectrum since the band at 1046 cm^{-1} appears strongly in the spectrum taken along the *a* axis. Other bands of the nitrate ion also obey the C_{2v} selection rules.

The ammonium ion bands show no polarization effects at all, in marked contrast to the pronounced polarization observed with the numerous bands attributable to the nitrate ion. This forcefully suggests that the ammonium ions are disoriented in the crystal structure, not necessarily by free or nearly free rotation, but perhaps by way of random orientation in the various unit cells.

Guanidinium ion has been investigated by Angell and others

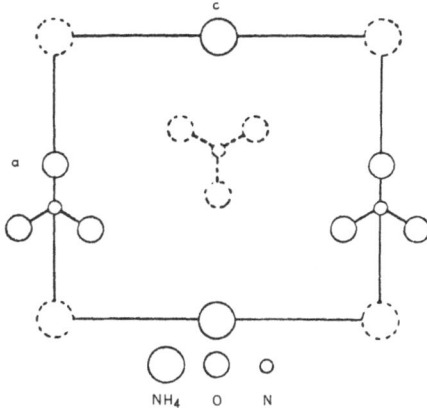

Fig. 6. Unit cell of NH_4NO_3 (IV) projected onto (010) plane. [Reprinted by permission from *J. Chem. Phys.* 18:1281 (1950).]

[5]. The aim of this work was to obtain reliable infrared data for the guanidinium ion, to consider its structure in the light of these results, and to calculate the normal vibrations of the ion.

This work included taking spectra of oriented crystals in a reflecting microscope. Guanidinium iodide crystallizes in the hexagonal system with space group C_{6v}^4, planar guanidinium ions all lying in the basal plane normal to the hexagonal axis [57]. However, as Fig. 7 indicates, the site symmetry for the ions is only C_{3v}. A small single crystal of guanidinium iodide was obtained by evaporating an acetone or alcohol solution, and was located in a reflecting microscope so that the sixfold axis of the crystal was parallel to the direction of incidence of the radiation. Bands that are forbidden in the D_{3h} symmetry are allowed by the site group C_{3v}, and these bands actually appear, if only very weakly, in the infrared spectra of polycrystalline samples. For example, the fully symmetrical stretching vibration at 1005 cm⁻¹ appears in the spectrum of guanidinium iodide as a very weak band. From the interpretation of the spectra it becomes clear that the selection rules appropriate to the C_{3v} site symmetry in the crystal of the iodide account very well for the observed infrared spectrum. Therefore it seems probable that the free ion has symmetry D_{3h}, and that the extra bands are caused by distortion due to crystal forces.

The study of a single crystal of ethylene by Brecher and

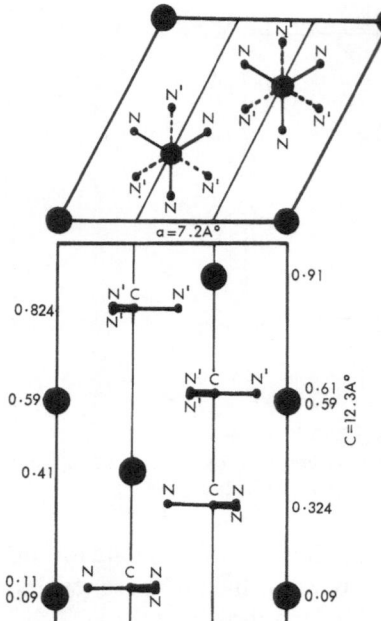

Fig. 7. The crystal structure of guanidinium iodide. Distances in fractions of unit cell lengths.

Halford [14] is an outstanding piece of work. Not only was the complete assignment of the absorption bands in the infrared obtained, but also the actual crystal structure (the space group) was determined through polarized infrared studies.

Three single crystals of ethylene were grown at liquid-nitrogen temperature and investigated spectroscopically. From an examination with a polarizing microscope it was found that each of the three samples contained the b axis. The comparison between ethylene vapor, liquid, and polycrystalline solid showed that the bands present in the vapor are shifted among the three phases only slightly. When a single crystal of ethylene was examined it was found that all bands showed substantial absorption in each polarization direction, indicating that no transition moment is perpendicular to a crystal extinction axis. It was also observed that the three infrared-inactive vibrations ν_2, ν_3, and ν_6 appear in the infrared spectrum of the liquid, although they are definitely absent from the infrared spectra of the crystals.

Another important observation was that the crystal extinction axes, as determined by visible light, always yielded the maximum dichroic effect for all three single-crystal specimens. The directions of the perpendicular extinction axes were independent of frequency only if they coincided with crystallographically significant directions. This could only be true if at least one crystal axis was perpendicular to the plane of the other two. This argument leads to the conclusion that the ethylene crystal belongs to either the orthorhombic or the monoclinic system.

Since the crystal spectra did not show the infrared-inactive vibrations, the site must have an inversion symmetry C_i. Since ethylene has a molecular symmetry D_{2h} the only subgroups of the molecular point group D_{2h} which contain a center of inversion are C_{2h} and C_i, and these are the only ones which can be considered further. This limits the choice of possible factor groups to only two: C_{2h} monoclinic and D_{2h} orthorhombic, since these are the only monoclinic or orthorhombic factor groups which contain sites having a center of inversion.

The observation that no bands polarize completely in any direction is used in deciding between the two possible site groups. Table I shows the mapping of the molecular point group D_{2h} through the site group C_{2h} into the factor group D_{2h}. This indicates that one of the infrared-active molecular species is allowed to appear only along one crystal axis, along which the other two species are forbidden. This is contrary to experimental evidence. On the other hand, mapping through site group C_i shows that the appearance of all infrared-active molecular modes is allowed along all three crystal axes. Therefore it is concluded that the molecules are located at sites having C_i symmetry, and C_i only.

Since splitting of the bands is also observed in some cases, the unit cell must contain at least two molecules. This would be in agreement with either a C_{2h} monoclinic unit cell or a D_{2h} orthorhombic unit cell. At this stage the results of x-ray crystallography were used [16]. These located only the carbon atoms of the ethylene crystal. The structure with respect to the carbon atoms alone is orthorhombic D_{2h}^{12}. After applying all the previous deductions of the ethylene crystal Brecher and Halford came to the conclusion that the complete space group structure of the ethylene single crystal was that of C_{2h}^5. Therefore this method of infrared polarized spectra has uniquely determined the space group of the ethylene crystal, while the x-ray results left a number of possibilities open for the orientation of the hydrogen atoms.

TABLE I*
Correlation Mappings

Mapping through C_{2h} site		Molecular group	Mapping through C_i site	
Factor group	Site group		Site group	Factor group
D_{2h}	C_{2h}	D_{2h}	C_i	C_{2h}

Correlation diagram:

Left (Factor group D_{2h} → Site group C_{2h}):
- A_g, B_{1g} → A_g
- B_{2g}, B_{3g} → B_g
- A_u, $B_{1u}(z)$ → $A_u(z)$
- $B_{2u}(y)$, $B_{3u}(x)$ → $B_u(x,y)$

Center (Molecular group D_{2h}):
- A_g, B_{1g}, B_{2g}, B_{3g}
- A_u, $B_{1u}(z)$, $B_{2u}(y)$, $B_{3u}(x)$

Right (Molecular group D_{2h} → Site group C_i → Factor group C_{2h}):
- A_g, B_{1g}, B_{2g}, B_{3g} → A_g → A_g, B_g
- A_u, $B_{1u}(z)$, $B_{2u}(y)$, $B_{3u}(x)$ → $A_u(x,y,z)$ → $A_u(z)$, $B_u(x,y)$

*Reprinted by permission from *J. Chem. Phys.* 35:1116 (1961).

TRANSITION MOMENTS

So far we have assumed that the transition moment is along or perpendicular to some chemical bond in the molecule. However, this is not always true, as will be illustrated by several examples to follow.

The infrared spectra of N-acetylglycine have been studied by Newman and Badger [45]. The crystal of acetylglycine is monoclinic and belongs to the space group C_{2h}^5. Small crystals several millimeters in size were grown from aqueous solution; faces parallel to (100) were obtained by cleaving the crystals parallel to this plane and then hand polishing down to a thickness of approximately 10μ. It would be desirable to have a crystal showing a face perpendicular to the (100) plane; however, this was not available. As an alternative, crystals could be obtained with the faces parallel to (011). These crystals were studied in the reflecting microscope.

Of the two strong hydrogen stretching fundamentals expected in this spectrum, the N—H stretching vibration is ascribed to the sharp band at 3140 cm⁻¹. This band is quite narrow, indicating that the hydrogen bonding is probably fairly weak, as was also indicated by the long N—O distance of 3.03 Å. A study of the polarization behavior of this band showed that its projection on the (100) plane makes an angle of 45° with the c axis. The corresponding angle between the N—H transition moment and the N⋯O line is 23°.

In a recent study Donohue and Marsh [25] have determined the actual direction of the N—H bond. Figure 8 shows that the hydrogen atom is 7° off the N⋯O line, while the transition moment of the N—H stretching vibration makes an angle of 15° with the N—H bond direction. In other cases of secondary amides where the hydrogen position is not known, the N—H transition moment direction has been referred to the C=O bond direction. For N-acetylglycine this angle between N—H transition moment and C=O direction is found to be 10°.

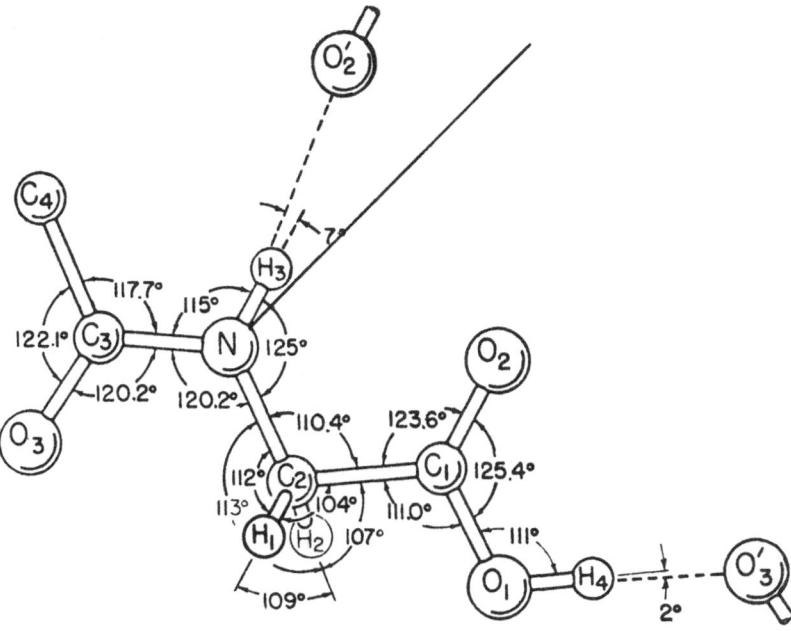

Fig. 8. The bond angles of N-acetylglycine.
[Reprinted by permission from *Acta. Cryst.* 15:944 (1962).]

The band ascribed to the O—H stretching vibration is of remarkably great breadth and of very high intensity. This strongly polarized band extends from 3000 to 1850 cm⁻¹ and appears with maximum intensity in the direction of the c axis of the crystal. Its strength and breadth indicate strong hydrogen bonding, which is in agreement with the short O—O distance of 2.56 Å, and its direction agrees very well with the direction of the O—O line as determined by the crystal structure.

Another example is provided by the study of the infrared spectrum and dichroism of acetanilide by Abbott and Elliott [1]. Acetanilide is orthorhombic, belongs to the space group D_{2h}^{15}, and has eight molecules to the unit cell. Samples were prepared by producing thin crystal layers of the compound by cooling from the melt between two plates of glass or rock salt. Some control of thickness was obtained by applying suitable pressure during cooling. The crystals obtained from the melt were plate-like, parallel to the (100) face, and spectra could be obtained from them with the electric vector along the b and c axes. A spectrum along the a axis was obtained by tilting the crystal around the b axis. However, a suitable specimen of a crystalline layer was found which was parallel to the (001) plane, and this made possible direct measurements with the electric vector along the a axis. These crystalline layers were studied in the reflecting microscope.

The N—H stretching region showed some very interesting effects. The main N—H stretching peak is at 3295 cm⁻¹ with the electric vector along the b axis, but the main peak for the a and c axial directions is at 3306 cm⁻¹ (see Fig. 9). The difference of 11 cm⁻¹ in the main N—H peaks is a consequence of a splitting of the band due to coupling between molecules. This is confirmed in striking fashion by the fact that when a 90% deuterated crystal was examined, the remaining N—H groups showed a single N—H vibration at the same frequency in all orientations. In such a material the remaining N—H groups are largely isolated from each other by N—D groups. Therefore, the coupling of N—H vibrations between neighboring molecules is destroyed, although all the molecules are in the same potential field as before. This technique of isotopic dilution has been successfully applied in other cases, for example, by Hiebert and Hornig on HCl [34]. It is interesting to note that the N—D vibration in the fully deuterated form also shows a factor group splitting which also is removed by isotopic dilution.

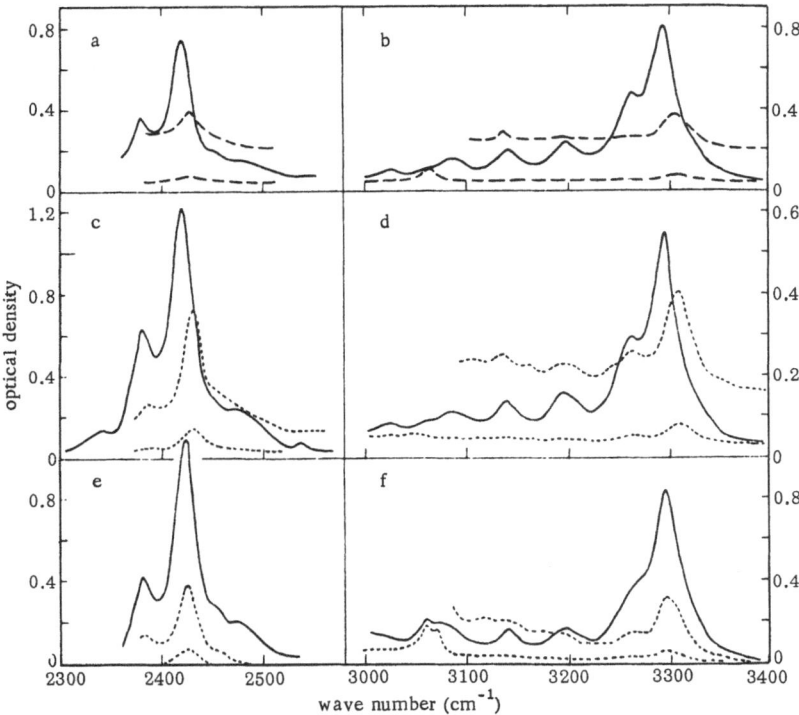

Fig. 9. N—H and N—D stretching modes in crystalline acetanilide. *E* vector parallel to *a* axis ---------, to *b* axis ———, to *c* axis · · · · · (a) N—D stretching mode (001) section; (b) N—H stretching mode (001) section; (c) N—D stretching mode (100) section; (d) N—H stretching mode (100) section; (e) N—D stretching mode (100) section; intensity νNH:νND *ca.* 12:1; (f) N—H stretching mode (100) section; intensity νND:νNH *ca.* 9:1. [Reprinted by permission from *Proc. Roy. Soc. (London)* A 234:252 (1956).]

as when the remaining N—D band is examined in 90% hydrogenated material.

The chief object of this work was the determination of the direction of transition moments and consideration of the relation between bond direction and transition moment. This latter consideration leads to interesting results in the amide group. It is found that the N—H stretching transition moment makes an angle of 12° with the CO bond direction.

The carbonyl stretching vibration, which according to Fraser and Price [31] could more properly be described as an out-of-phase

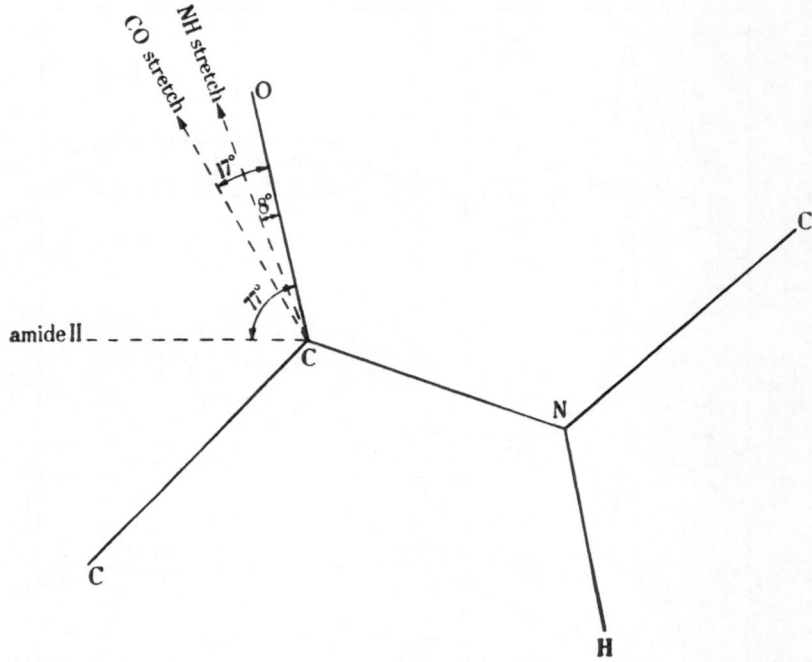

Fig. 10. Transition moment directions in N,N'-diacetylhexamethylenediamine. [Reprinted by permission from *Proc. Roy. Soc. (London)* A 249:167 (1958).]

mode of the C=O and C—N stretching vibrations, has a transition moment about 22° from the direction of the CO bond. Because of the mixed nature of this vibration, and also because of the electron flow during the vibration (orbital following), it is not unexpected that the transition moment would not be along the C=O bond, but would be inclined at an angle to that bond in the direction of the C—N bond (see Fig. 10).

The band at 1558 cm⁻¹ (sometimes referred to as the Amide II band) is thought to arise from accidental resonance between an N—H bending and an inplane NCO stretching vibration. Since this vibration is also mixed, it is reasonable to expect that its transition moment would not be exactly perpendicular to the N—H bond direction. Ambrose and Elliott's results on acetanilide indicate that this Amide II vibration makes an angle of +72° with the C=O bond direction.

The "amide" infrared bands are of great interest in the struc-

amide plane (1)   c axis   amide plane (2)

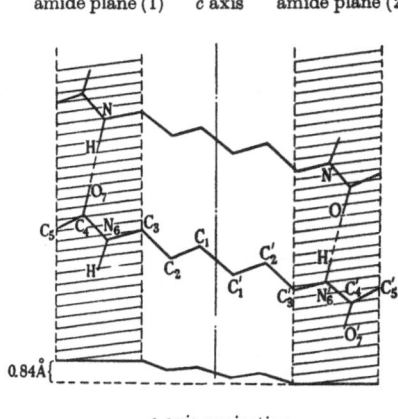

c axis projection

Fig. 11. Diagram showing two adjacent molecules in the N,N′-diacetylhexa-methylenediamine crystal. [Reprinted by permission from *Proc. Roy. Soc. (London)* A 232:106 (1955).]

tural study of proteins as well as of amides and polyamides. N,N′-diacetylhexamethylenediamine, studied by Sandeman [54], is a very good model compound for these studies because it can be obtained in a single crystal of triclinic form with one molecule per unit cell, and so yields a unique direction of dipole-moment change for each infrared band.

The crystal structure of this compound has been previously determined [6], and Fig. 11 gives the c axis projection of the molecule. The hexamethylene chain in the middle of the molecule is planar, while the two remaining portions of the molecule at each end of the chain lie on two parallel planes. These two planes could be called the amide planes and are inclined at an angle of 11°28′ to the hexamethylene plane. The compound crystallizes from the melt between rock salt plates with the (100) face in contact with the plates. The direction of the c axis is strongly marked by cleavage. The studies were made on polycrystalline oriented samples prepared from the powdered compound as compressed between rock salt plates, and melted over a Bunsen flame. The plates then were cooled in a temperature gradient, and crystallization occurred slowly along the sample.

In the interpretation of the spectra, deuteration proved to be of great help. Comparison of the spectra of deuterated and undeuterated samples enables the bands to be sorted into two groups, namely, those affected and those unaffected by deuteration. The bands unaffected by deuteration are presumably those belonging to the hexamethylene chain or to the methyl groups, while the bands affected by deuteration are those connected with the amide group.

Sandeman makes a number of assignments for the spectrum of the hexamethylene chain and obtains directions of transition moments which are in satisfactory agreement with previous assignments for long methylene chains.

Of the bands produced by the amide group, three are discussed in detail. One of these is the N—H stretching band at 3310 cm^{-1}, the polarization of which indicates that its transition moment makes an angle of 8° with the direction of the CO bond. The Amide I or carbonyl stretching vibration appears as a doublet, at 1632 and 1642 cm^{-1}. Calculations on the polarization indicate that the transition moment is distorted about 17° from the direction of the CO bond. The Amide II band also appears as a doublet, at 1540 and 1555 cm^{-1}. However, it was shown by Beer [8] that if a double peak is obtained with polarized light, leading to two apparent transition moments, then the true transition moment direction can be calculated. The result of this calculation shows that the Amide II band has a transition moment which makes an angle of 77° with the carbonyl bond. It can be seen that the results on acetanilide and N,N'-diacetylhexamethylenediamine are in good agreement as far as the transition moments of the amide group are concerned. The application of these results to protein spectra and to the evaluation of protein structures will be discussed in a later section.

An interesting example of transition-moment determination from polarized infrared spectra in a compound not of the amide type is the study on azobenzene by Daasch [23]. This compound was chosen for study because the phenyl group has strong tendencies toward group frequencies, and most of its vibrations have been well characterized. The purposes of the study were to determine the directions of the transition moments of various vibrations of the phenyl group, and to find out whether these directions are constant for a given class of vibrations.

The structure of the *trans* isomer of azobenzene has been determined. It belongs to the monoclinic system with space group symmetry C_{2h}^5, and has four molecules per unit cell [24]. The molecules have symmetry C_i. Oriented crystals were prepared by the evaporation of solutions or by the solidification of melts between two sodium chloride plates. Sample thickness could be controlled by pressure applied to the plates while the sample was in the liquid state. It was noted that the samples contained two different crystal orientations. One of these specimens was oriented with the *a* and *c* axes of the monoclinic cell in the plane of the sample. The other had the *a* and *b* axes in the plane of the specimen, with the *a* axis the direction of greatest growth. It was fortunate that these two orientations were available, since it is necessary to have the spectra of the crystal in two orientations in order to calculate the direction of the transition moment.

The group vibrations of the phenyl group can be divided into three classes according to direction: One direction is the line joining carbon atoms 1 and 4; the second direction is the line joining carbon atoms 2 and 6; and the third direction is perpendicular to the phenyl ring. From the study of the polarized spectra along the three crystallographic axes, Daasch has determined the transition moments of a large number of bands and has found that within a given symmetry class of vibrations the transition-moment direction is not constant, as ideally it should be. The variation is not so great, however, that a number of bands cannot be grouped in their correct symmetry classes by knowing only the directions of the transition moments. The observed directions of the transition moments were as much as 6 to 24° removed from the directions predicted on the basis of isolated phenyl ring vibrations.

Daasch carried out an analysis of the data considering no interaction between the molecules in the unit cell (that is, according to the oriented gas model), and this explains the polarization data with moderate success. A factor group analysis, which considers the vibrations as arising from closely coupled molecules in the unit cell, does not correlate well with the data. However, this disagreement could be because of a number of unresolved bands which are closely overlapped in the spectrum.

POLYMERIC MATERIALS

Infrared analysis has been such a powerful tool in exploring the structure of smaller molecules that there can be no doubt about the desirability of applying the method to large molecules. Sutherland and his co-workers in a series of articles [37, 39, 40] have described the basis of the interpretation of infrared spectra of high-molecular-weight polymers.

In the first of these papers they discussed the experimental methods employed in these investigations and the basis for the theoretical methods that are used in interpreting the spectra of polymers in general. To develop the selection rules for polymer spectra, the following assumptions are made: that the polymer molecule is infinitely long, that these molecules have a regular arrangement in a crystalline lattice, and that the interaction forces between polymer chains are so small that the magnitudes of the fundamentals are not greatly changed by this interaction. This leads to the consideration of the spectrum of a single polymer molecule. The differences between the spectrum predicted for a single polymer molecule and that predicted for a true unit cell will be of two kinds. First, certain of the fundamentals of a single polymer molecule will be split into several components in the cell, depending on the interaction forces between molecules in the unit cell. Second, certain modes which are totally inactive for the single polymer molecule may become active in the unit cell when the site symmetry of the molecule in the unit cell is lower than the symmetry of the isolated molecule.

The selection rules derived apply to the crystalline polymer. The spectrum of the amorphous form may differ from this by the appearance of additional absorption bands. These may be due either to breakdown of selection rules operating in the crystal or to the existence of different rotational isomers in amorphous form.

The selection rules derived by such methods will predict the dichroic properties of the vibrations active in the infrared. However, to predict the actual frequencies of the expected vibrations, a number of other methods must be called on. Sutherland and his co-w ·kers suggest useful application of the following: approximate noi mal coordinate analysis; group frequencies, many of which are well established (this is especially true of hydrogenic stretching vibrations, of C—H hydrogenic deformation frequencies, of carbon-halogen stretching frequencies, and of the stretching frequencies of double or triple bonds between carbon, nitrogen, and oxygen); substitution of deuterium for hydrogen, for identifying hydrogenic frequencies; and chemical substitution of groups. A theoretical method is also given for the calculation of skeletal frequencies of long-chain molecules. This is applied to an infinitely long, planar, zigzag chain containing two or four carbon atoms as units. Frequencies for the active skeletal vibrations expected from such a chain have been listed.

In the second paper of this series these results are applied to the interpretation of the infrared spectra of polyethylene. Orientation in polyethylene was produced by cold-drawing as fully as possible. It is known that this treatment produces orientation in which the crystal or chain axes become parallel to the stretching direction. In addition, partial orientation of the chains occurs in the amorphous regions. Some long-chain n-paraffin hydrocarbons were used for comparison to polyethylene, and these were obtained as single crystals by evaporation of a solution. It is known from x-ray diffraction studies that polyethylene in the crystal arrangement contains chains in the planar zigzag configuration, and that there are two chains in a unit cell.

First a single polyethylene chain is considered, the repeating unit of which is taken as consisting of two CH_2 groups. The factor group for a single polyethylene chain is found to be D_{2h}. Application of this group shows that modes of species B_{1u} and B_{2u} are expected to show perpendicular polarization with respect to the chain axis, and that modes under B_{3u} should have parallel polarization. Also, the mutual exclusion rule between infrared and Raman spectra holds. The derivation of the normal modes of the chain can be done by a comparison with the normal modes of the component CH_2 groups. The nature of the infrared-active vibrations of a single polyethylene chain is thus derived. Assignment of a number of infrared-active CH_2 fundamentals to the polyethylene spectrum is carried out by making use of the polarization data (see Table II).

After this, the predicted spectrum of crystalline polyethylene is considered, since the complete spectrum of the crystal can be only partially predicted by consideration of a single chain. The unit cell structure of polyethylene belongs to space group D_{2h}. For the infrared-active vibrations of species B_{1u}, B_{2u}, and B_{3u} this predicts the same polarizations as were predicted in the above consideration of a single chain. In the spectrum of the crystal, splitting of some of the fundamental frequencies is expected. However, interactions which lead to a splitting of frequencies in a crystal will not drastically change the positions of the fundamental frequencies. In the polyethylene spectrum one such splitting is the much discussed doubling of the band at about 725 cm^{-1}. In a study of the two components of the 725 cm^{-1} band in the spectra of monoclinic $C_{36}H_{74}$, complete agreement with the predicted polarization is obtained, the 721 cm^{-1} band being polarized along the b

TABLE II
Infrared Spectrum and Assignments for Polyethylene

Frequency, cm⁻¹	Intensity	Polarization	Assignment* Reference [37]	Reference [36]
2959	w(sh)	···	$\nu_a(CH_3)$	$\nu_a(CH_3)$
2924	vs	σ_a	} $\nu_a(CH_2)$	$\nu_a(CH_2)\,(B_{1u})$
2899	vs	σ_b		$\nu_a(CH_2)\,(B_{2u})$
2874	vw	···	$\nu_s(CH_3)$	$\nu_s(CH_3)$
2857	vs	σ_b	} $\nu_s(CH_2)$	$\nu_s(CH_2)\,(B_{2u})$
2850	vs	σ_a		$\nu_s(CH_2)\,(B_{1u})$
2640	w	σ		1168 + 1473
2295	vw	σ		1176 + 1131
2130	vvw	σ		
2010	vw	π	1303 + 721	1303 + 721
1890	vw	σ		1168 + 731
1805	vvw	π		
1710	vw	σ		
1470	s	σ_a	} $\delta(CH_2)$	$\delta(CH_2)\,(B_{1u})$
1460	s	σ_b		$\delta(CH_2)\,(B_{2u})$
1456	vw	···	$\delta_a(CH_3)$	$\delta_a(CH_3)$
1375	m	$\pi(?)$	$\delta_s(CH_3)$	$\delta_s(CH_3)$
1369	w	π	$\gamma_w(CH_2)$	
1353	w	π	} $\gamma_w(CH_2)$ amorphous	$\gamma_w(CH_2)$ amorphous
1303	w	π		
1176	vvw	π		$\gamma_w(CH_2)\,(B_{3u})$
1150	vvw	π		
1110	vvw	π		
1080	vw	σ	$\nu(C-C)$ amorphous	$\nu(C-C)$ amorphous
1065	vw(sh)	σ	$\nu(C-C)$	skeletal
965	vvw	σ		$\gamma_r(CH_2)$ amorphous
888	vw	σ	$\gamma_r(CH_3)$	$\gamma_r(CH_3)$
731	s	σ_a	} $\gamma_r(CH_2)$	$\gamma_r(CH_2)\,(B_{1u})$
721	s	σ_b		$\gamma_r(CH_2)\,(B_{2u})$
600	vw	···		
543	w	···		
200	vw	···		

*ν = stretching, δ = bending, γ_w = wagging, γ_r = rocking.

axis and the 731 cm^{-1} band along the *a* axis. Further evidence that the splitting of the 725 cm^{-1} band arises from an interaction between chains in the crystal is provided by the disappearance of this splitting in a study of solid solutions of a paraffin in a deuterated paraffin. This effect of isotopic dilution has been discussed in a previous section.

The two articles described here are by no means the last dealing with the vibrational spectrum of polyethylene. A considerable amount of work has been published since, both on experimental (spectrum of the overtone region [48], Raman spectrum [15, 48]) and theoretical aspects [41, 56]. This has resulted in some modification of the vibrational assignment of polyethylene and a nearly complete understanding of all the normal modes; a recent assignment due to Krimm [36] is included in Table II.

Actually, theoretical articles dealing with the vibrations of the polyethylene chain seem to appear with such regularity that there is little doubt that a few more will have been published before this present article appears in print.

The methods discussed above for polyethylene have also been applied to the interpretation of the polarized infrared spectra of a number of other oriented polymers such as polytetrafluoroethylene [40], polyvinyl chloride [36], polyvinyl alcohol [36], polystyrene [36], polyformaldehyde [49], polychloral [50], and polyethylene glycol [43], to mention a few.

The proteins are an interesting class of polymeric molecules. The infrared spectra of all proteins resemble each other very closely. This is not surprising since all protein molecules are built from polypeptide chains which contain varying amounts of some 20 amino acid residues. Despite the resemblance of protein spectra, considerable progress has been made in establishing relationships between the structures of protein molecules and their infrared spectra. Many papers deal with the infrared spectra of proteins and polypeptides, and in this review we would like only to talk about an article by Beer, Sutherland, Tanner, and Wood [9] which summarizes the results obtained with protein molecules.

These authors divide the proteins studied into class A, in which the polypeptide chain is coiled or folded, and class B, in which the peptide chain is believed to be essentially extended. The spectra of the proteins in class A are very similar, and it is unlikely that one member of this group can be distinguished from another purely

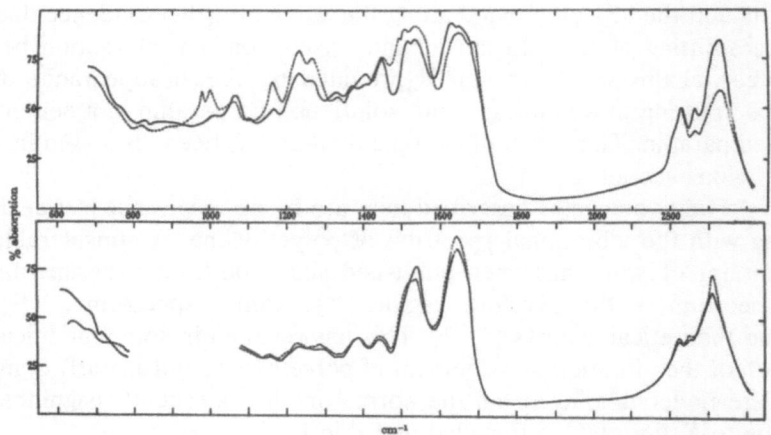

Fig. 12. Infrared spectra of silkworm gut (top) and elephant hair (bottom)
. . . . . . electric vector parallel, ——— electric vector perpendicular to fiber axis.
[Reprinted by permission from *Proc. Roy. Sog. (London)* A 249:159 (1958).]

by its infrared spectrum. However, the spectra of proteins of class B
are sufficiently different from those of class A and from one another
to make spectroscopic differentiation possible in a few cases.

In many synthetic polypeptides as well as many proteins, the
structure is that of the α-helix, in which the $CO \cdots HN$ hydrogen
bonds are nearly parallel to the axis of the helix. A different structure also frequently occurring is the fully extended configuration
with the $CO \cdots HN$ hydrogen bonds nearly perpendicular to
the axis of the peptide chain. This structure is referred to as the
β-structure.

It has been reported by several authors that these two forms
can be differentiated by the frequencies of the Amide I band near
1650 cm^{-1} and the Amide II band near 1550 cm^{-1}. However,
Beer *et al.* show that this frequency criterion established for the
α and β forms in the synthetic polypeptides can probably not be
used to give a reliable indication of such chains in proteins. In
this case, infrared dichroic measurements are a much more important criterion to distinguish the two structures. Beer *et al.* list
the bands found for the α and β forms. In the β forms the N—H
stretching vibration as well as the CO stretching vibration is perpendicular to the direction of the chain, while the Amide II band
is parallel. The exact opposite occurs in the case of the α forms:

The N—H stretching and CO stretching are parallel, while the Amide II band is perpendicular to the chain direction. Figure 12 shows the polarized infrared spectra of silkworm gut and elephant hair; from the polarizations, silkworm gut is clearly of the β form while elephant hair is of the α form.

One objective is to identify the bands arising from the vibrations of the peptide N—H groups. In this, deuteration studies have been of great help. The hydrogens in the N—H bond of the peptide link are easily replaced with deuterium by exchange in heavy water. When the deuterated spectra are studied it is realized that quite a large number of bands disappear from the hydrogen-containing spectrum, while a number of new bands appear in the deuterated spectrum. This indicates that the NH vibrations are sometimes quite strongly coupled with other vibrations of the molecule. In this connection it is interesting to note that the position of the CO stretching band in proteins is definitely shifted to longer wavelength by deuteration. The change in frequency is about 12 cm⁻¹. The importance of this observation is that the CO frequency is not entirely separable and independent of the motion of the hydrogen atom in the NH group.

In this study the only deuterated protein in which dichroic effects could be observed was silkworm gut. Most remarkable was the great change on deuteration in the dichroism of the NH band at 3190 cm⁻¹, as illustrated in Fig. 13. The dichroic ratio has de-

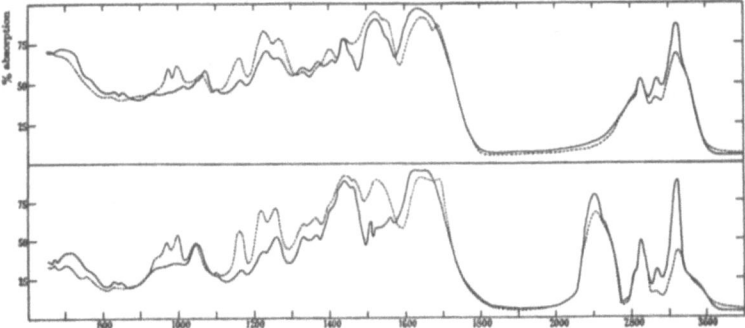

Fig. 13. Effect of deuteration on the dichroic infrared spectra of silkworm gut. Top, natural; bottom, deuterated; · · · · · electric vector parallel, —— electric vector perpendicular to the fiber axis. [Reprinted by permission from *Proc. Roy. Soc. (London)* A 249:161 (1958).]

creased from 0.54 in the undeuterated protein to 0.25 in the deuterated silkworm gut. The 1535 cm^{-1} band also shows a definite increase of dichroism on deuteration. This seems to indicate that deuteration occurs preferentially on the less well-oriented parts of the polypeptide chain in silk. This phenomenon has also been reported for other proteins, and illustrates the danger of trying to make deductions on the correctness of dichroic measurements of any absorption band without taking into account the possible misalignment of molecules in these proteins.

These deuteration experiments have been of considerable help in evaluating the protein structure, and have shown interaction between NH vibrations and other vibrations of the amide group in many cases.

The observed dichroism of a band in a polymer depends upon two factors: the direction of the transition moment vector of the repeating unit relative to the axis of the polymer and the degree to which the axis of the molecules of the polymer are aligned parallel to one another in the sample. Several articles in the literature [7, 28-30] deal with the problem of determining the effect of the unoriented part of the sample on the dichroic ratios.

Beer [7] has shown that if the directions of three linearly independent transition moments are known relative to the repeating unit of the polymer, then from the dichroic ratios of the bands it is possible to find both the degree of alignment of polymers and the orientation of the repeating unit relative to the axis of the polymer. This method depends upon several assumptions as discussed in the article by Beer *et al.*

Studies on simple amides have shown that the characteristic amide bands change very little with changes of groups which are attached to the peptide link. Therefore it seems reasonable to assume that transition moments determined on small amide molecules can be applied in the case of polypeptides and protein molecules. From this assumption it is possible to calculate the direction of the transition moment vector for various characteristic vibrations of the peptide link for any model of a protein.

With these directions of transition moment vectors, together with experimentally observed dichroic ratios for the corresponding bands, it is possible to use the method of Beer to find from each band a value of the disorientation parameter F for any given sample of protein. Since F is characteristic of the sample only, only those structures can be considered quantitatively consistent with this kind of analysis which yield approximately the same F value for all bands observed. In the case of proteins, the bands used are usually the four amide bands.

Several examples of such calculations are given in the article by Beer *et al.* For example, for both porcupine quill and elephant hair, it is found that the structure consistent with the infrared dichroism is the α-helix. Similarly, in the case of silkworm gut, best agreement is obtained for the open chain or β-structure. For feather keratin the best agreement is again with the fully extended β-structure. However, another structure has also been proposed for feather keratin [38], and it is found that it is not possible to discriminate between the two structures on the basis of infrared dichroic results alone.

Three structures have been proposed for collagen. Of these, the structure proposed by Ramachandran and Kartha shows the best agreement with the calculation of dichroism previously described. The agreement with the structure proposed by Rich and Crick is not nearly as good, but this structure cannot be rejected on the basis of infrared data only. On the other hand, the structure proposed by Pauling and Corey is definitely unacceptable.

Another interesting long-chain molecule with repeating units is the sodium salt of deoxyribonucleic acid (DNA). A difficulty arises in the infrared study of this compound because the crystallinity of DNA is very sensitive to the humidity of the surrounding air. Therefore, the infrared spectra have to be obtained on oriented films under conditions of controlled humidity. Sutherland and Tsuboi [55] have devised a special cell for this purpose (see Fig. 14). The sample is deposited on the inside of one of the AgCl windows, and the humidity is controlled by placing saturated solutions of various salts with excess of the solids at the bottom of the cell. When the cell is evacuated, the sample can be dried out, and when a heavy-water solution is placed in the bottom, vapor-phase deuteration of the sample is achieved. Thus the same sample with the same orientation can be examined in both the protonated and deuterated forms.

DNA is made up of four heterocyclic bases (adenine, guanine, cytosine, and thymine) held together by chains made of deoxyribose rings connected by phosphate ester linkages. The general structure of this compound, as deduced by Watson and Crick [22], is

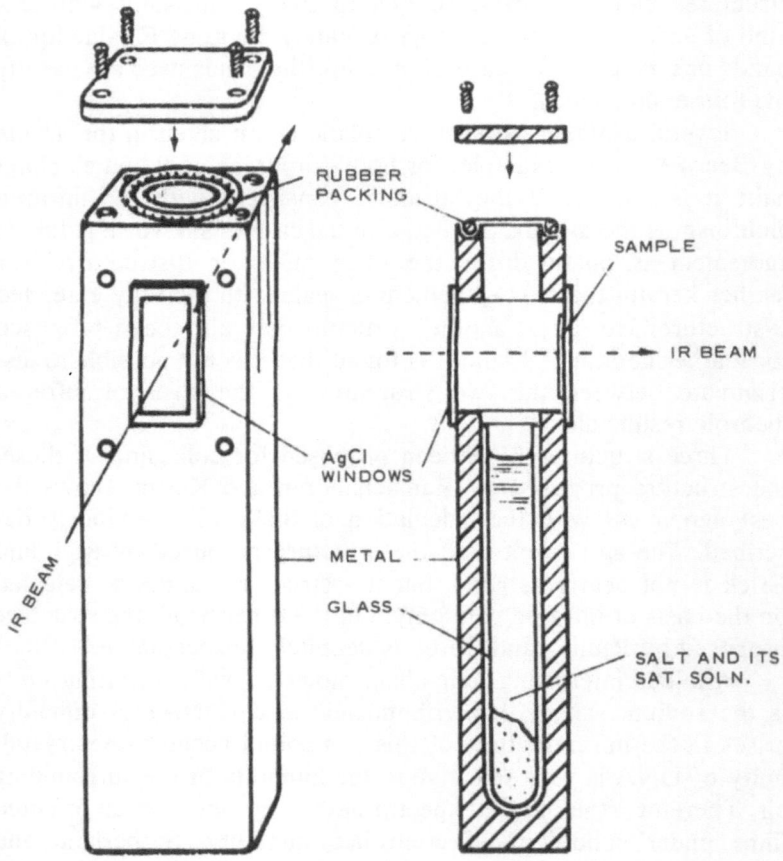

Fig. 14. Absorption cell in which the humidity can be controlled.
[Reprinted by permission from *Proc. Roy. Soc. (London)* A 239:448 (1957).]

a double helix with the strands held together by hydrogen bonds
between the bases. This complicated structure would be expected
to give a very complicated infrared spectrum, and hardly any interpretation could be expected. On the other hand, it is a fact that important results came out of the study of polarized infrared spectra
as far as the orientation of the groups in the helical chain is concerned.

Oriented films of DNA can be prepared by making a viscous
solution in water and then spreading this with a spatula on a silver
chloride plate. Spreading is done along the length of the plate so

that orientation of the fibers parallel to the length of the plate is achieved. Under the polarizing microscope such a film shows quite strong negative birefringence. At 90% relative humidity the estimated water content of the DNA film is between 45 and 50% of dry weight.

Figure 15 shows the polarized infrared spectrum of sodium DNA at various relative humidities. The changes in the spectrum caused by the humidity are very striking. At higher humidities the film shows much stronger dichroism, indicating that it is more ordered at high humidity. All these spectral changes are reversible, at least qualitatively.

Sutherland and Tsuboi have shown that a spectrum similar to the spectrum of DNA can be built up by adding the spectra of the constituent bases [4], phosphate groups, and sugar units. In this

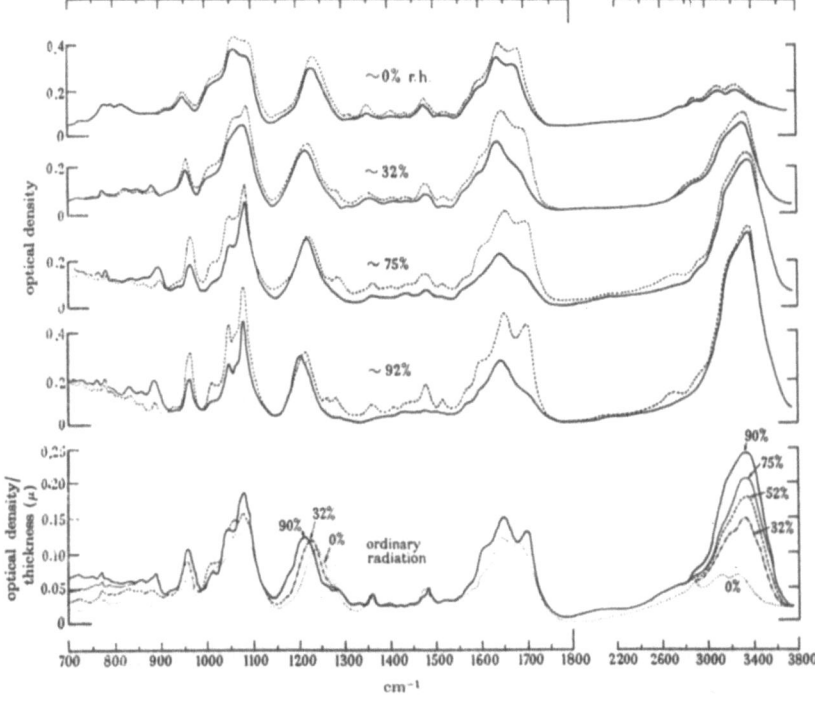

Fig. 15. The infrared spectrum of NaDNA at various relative humidities. ———— electric vector parallel, ... perpendicular to the orientation direction. [Reprinted by permission from *Proc. Roy. Soc. (London)* A 239:450 (1957).]

way a fairly satisfactory understanding of the DNA spectrum can be achieved.

In the 3-μ region NH_2 bands are observed which show polarization perpendicular to the fiber axis. Since all the NH_2 bonds are in the plane of the heterocyclic rings, this result indicates that the ring planes are arranged perpendicular to the fiber axis. At high humidities this region is obscured by a strong water band which does not show any dichroism.

The $1600-1750$ cm^{-1} region is characteristic mostly of the in-plane vibrations of the base residues, consisting of C=O, C=N, and C=C bands. It will be seen from the spectrum that this region always exhibits polarization perpendicular to the chain direction. This indicates that the planes of the rings are perpendicular to the chain. Water at 1660 cm^{-1} interferes in this region also, but when deuterated DNA is examined, the water band is shifted to 1225 cm^{-1}, and very strong perpendicular dichroism is revealed.

Since we are dealing with the sodium salt of DNA, its phosphate part is regarded as having the structure

$$\left[\begin{array}{c} -O \\ -O \end{array} \!\! \begin{array}{c} O \\ \diagup P \diagdown \\ O \end{array} \right]^{-}$$

To establish the frequencies of vibrations associated with this group, simpler compounds were used for comparison: the hypophosphite anion $H_2PO_2^-$ and the dimethylphosphate anion $(CH_3O)_2PO_2^-$. Each of these compounds has a strong band in the $1180-1240$ cm^{-1} region that can be assigned to the PO_2^- antisymmetric stretching vibration. Therefore, the isolated strong band at 1230 cm^{-1} in DNA is assigned to the antisymmetric stretching motion of the PO_2^- ion. The PO_2^- antisymmetric stretching vibration has its transition moment along the direction of the O···O line.

The PO_2^- symmetric stretching vibration is found at 1048 cm^{-1} in the hypophosphite anion, and at 1054 cm^{-1} in the dimethylphosphate anion. Hence the band at 1050 cm^{-1} in wet sodium DNA is assigned as a PO_2^- symmetric stretching vibration. This band shows very strong perpendicular dichroism and its transition moment is expected to be in the direction of a line bisecting the OPO angle.

From the dichroism of the bands of the PO_2^- group a fairly definite suggestion is derived about the orientation of this group. The
$O \cdots O$ line is inclined at an angle of about 60° from the helical
axis, and the bisector of the angle OPO is directed almost perpendicular to the helical axis. In the structure as suggested by
Crick and Watson, the phosphate group orientation is such that the
$O \cdots O$ line is perpendicular to the helical axis and that the bisector
of the angle OPO is almost parallel to the fiber axis. This orientation
should give rise to perpendicular dichroism in the antisymmetric
stretching vibration and parallel dichroism in the band assigned to
the PO_2^- symmetric stretching vibration. The infrared results are in
direct contradiction to this prediction.

While the work of Sutherland and Tsuboi was in progress, a
modified model structure of sodium DNA based on more recent
x-ray work was published [27]. In this new structure the orientation
of the PO_2^- group is in substantial agreement with the orientation suggested by the polarized infrared spectral studies.

As another example of the use of polarized infrared spectra in
the determination of the structures of polymer molecules, we
would like to quote the work of Tsuboi [58] on cellulose. The experiments were carried out on natural flax fibers arranged com-
pactly side by side so that all the fiber axes were parallel.

In the infrared spectra strong dichroism was observed in almost
the whole range of the spectrum. For some bands, the dichroic
ratios were nearly zero or infinite. This fact indicates that the
sample was in a well-oriented state, so that the results can be discussed on the basis of the crystal structure of cellulose.

The accepted model of the crystal of cellulose is the one proposed by Meyer and Misch [42]. In this model the unit cell is monoclinic and contains two cellobiose residues, i.e., four glucose
residues. If one assumes tetrahedral bond angles, then the directions of the C—H bonds are automatically determined. One of the
features of this structure is that all the tertiary C—H bonds are
directed perpendicular to the fiber axis. The CH_2 groups are oriented in such a way that the bands due to the CH_2 symmetric and
antisymmetric stretching vibrations should be polarized parallel
and perpendicular to the fiber axis, respectively.

Figure 16 shows the region from 2700 to 3100 cm⁻¹ in the infrared spectrum of flax fiber. The strong band observed at 2907
cm⁻¹, polarized perpendicular to the fiber axis, is most likely due

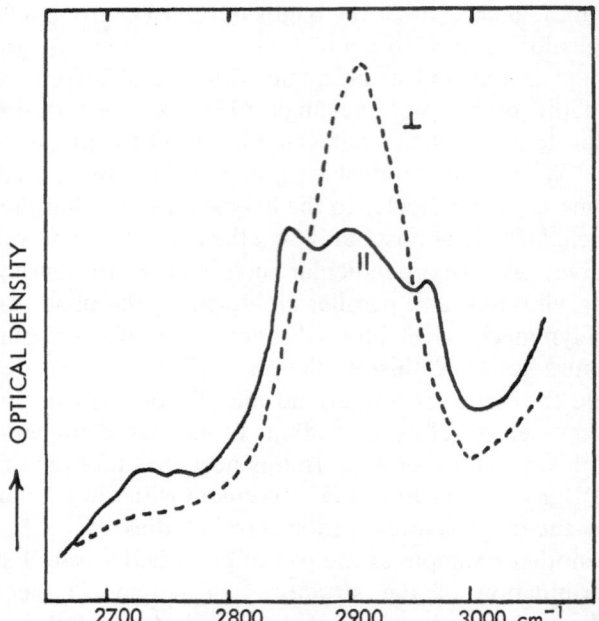

Fig. 16. Infrared spectrum of flax fibers in the 2700–3100
cm⁻¹ region. [Reprinted by permission from *J. Polym. Sci.*
25:165 (1957).]

to the C—H stretching vibration. Besides this, there are two distinct peaks at 2851 and 2967 cm⁻¹, both of which are polarized
parallel to the fiber axis. These would be assigned to the CH_2
symmetric and antisymmetric stretching vibrations, except that
their dichroic properties do not agree with the properties predicted
from the crystal structure. However, Tsuboi suggests in this paper
that the structure can be rearranged at the C_5 and C_6 of the glucose
residue so as to bring the orientation of the CH_2 groups into agreement with the observed polarizations.

It was also found that the amorphous part of cellulose was
much more easily deuterated than the crystalline part. This effect
was used in the experiment to study the arrangement of the OH
groups from the dichroism of the 3000–3600 cm⁻¹ region. In
order to get rid of the unoriented amorphous part of the sample a
bundle of ramie fibers was immersed in sodium deuteroxide solution,
washed with heavy water, and dried without contact with water
vapor. This treatment exchanged hydrogen for deuterium in both

the crystalline and amorphous regions. When this sample was exposed to atmospheric water vapor the amorphous part of the compound was rehydrogenated while the oriented crystalline part of the sample was left deuterated.

The infrared spectrum of the O—D stretching region is shown in Fig. 17. Of the six O—D bands observed, five clearly have parallel dichroism while the one at 2527 cm⁻¹ has perpendicular dichroism. Of course, from this spectrum it is not possible to say which of the O—H or O—D groups corresponds to this band. However, it can be concluded that a small fraction of the O—H bonds are perpendicular to the fiber axis, while most of them are parallel to this axis.

STREAMING DICHROISM

In the introduction it was mentioned that the study of polarized infrared spectroscopy is always the study of spectra of solids. An

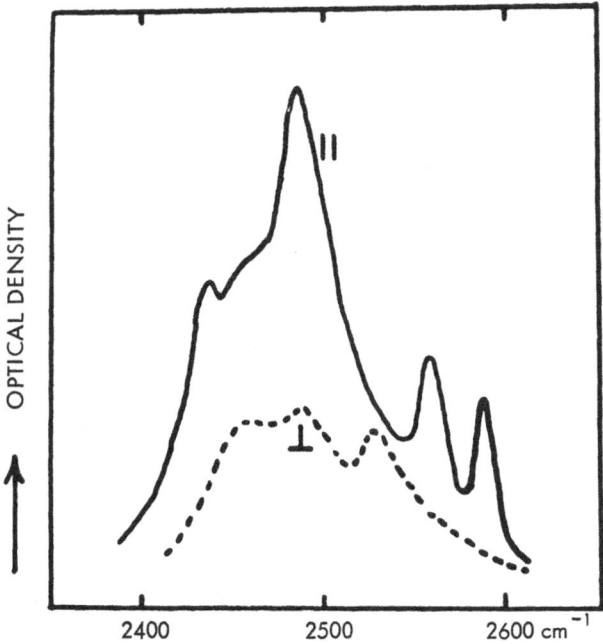

Fig. 17. Infrared spectrum of NaOD-swelled ramie fiber in the O—D stretching region. [Reprinted by permission from *J. Polym. Sci.* 25:168 (1957).]

exception is the work by Bird and Blout on infrared streaming dichroism of polymer solutions [11]. This study extends the method of infrared dichroism to polymers in solution. It is important because many interesting polymers normally exist only in solution, or fail to crystallize, or are greatly altered on deposition as a film.

The cell used is a modification of a standard infrared liquid cell, and orientation of the polymer in solution is achieved through a velocity gradient caused by running the liquid rapidly through the cell. Since the thickness is very small, a small rate of flow through the cell produces very large gradients. In this case the direction of flow is perpendicular and the velocity gradient is parallel to the beam, whereas in ordinary birefringence experiments the gradient is perpendicular to the beam. The equipment necessary for producing such a flow through an infrared cell is described in an article by Bird, Parrish, and Blout [12].

The attractive feature of this method measurement is that one can observe ordinary transmittance without flow, then start the flow and observe the alteration produced by orienting the macromolecules in solution. An increase or decrease of a certain absorption band indicates that the polymer has achieved orientation in the solution, and the band being observed has dichroism.

The calculation of the velocity gradient in such a system as well as calculation of the observed optical density change as a function of reduced gradient has been outlined by Bird [10]. Experiments were carried out on poly-γ-benzyl-L-glutamate in a solution of chloroform plus $\frac{1}{2}\%$ of formamide. The 3300 cm^{-1} band rather closely followed the theoretical density changes calculated for an ideal polymer with a perfectly oriented parallel absorber. Therefore, it can be considered that the dipole responsible for this band is perfectly oriented along the polymer helix.

Similarly, the 1550 cm^{-1} band (Amide II) showed perpendicular dichroism corresponding to perfect perpendicular orientation of the responsible dipole relative to the axis of the helix. On the other hand, changes observed for the 1650 cm^{-1} band (Amide I) are smaller but of the same sign as those for the 3300 cm^{-1} band, so that the Amide I band arises from an imperfectly oriented parallel absorber. The benzyl ester C=O stretching band at 1730 cm^{-1} arises from a group on a flexible side chain and shows no dichroism in the experiment. These observations have been compared with other observations on oriented films, and very clearly confirm the helical structure of this molecule in solution.

Further interesting results were obtained when spectra observed with poly-L-glutamic acid and its sodium salt were compared. The free acid in dioxane – heavy-water solution gave results which were practically identical to those on poly-γ-benzyl-L-glutamate: parallel dichroism for the 3300 and 1650 cm^{-1} bands, and perpendicular dichroism for the 1550 cm^{-1} band. On the other hand, the sodium salt in heavy-water solution showed no significant dichroism with gradients of the same magnitude as those used in the free acid experiment. The failure to observe dichroism in this case is consistent with the view that electrostatic repulsion destroys the helical configuration. The streaming dichroism results clearly point to a random configuration.

REFERENCES

1. Abbott, N. B., and Elliott, A., *Proc. Roy. Soc. (London)*, A 234:247 (1956).
2. Ambrose, E. J., Elliott, A., and Temple, R. B., *Proc. Roy. Soc. (London)*, A 206: 192 (1951).
3. Angell, C. L., Ph.D. Thesis, Cambridge, 1955.
4. Angell, C. L., *J. Chem. Soc.* p. 504, (1961).
5. Angell, C. L., Sheppard, N., Yamaguchi, A., Shimanouchi, T., Miyazawa, T., and Mizushima, S., *Trans. Faraday Soc.* 53:589 (1957).
6. Bailey, M., Acta Cryst. 8:575 (1955).
7. Beer, M., *Proc. Roy. Soc. (London)*, A 236:136 (1956).
8. Beer, M., Symposium on Molecular Structure and Spectroscopy, Columbus, 1956.
9. Beer, M., Sutherland, G. B. B. M., Tanner, K. N., and Wood, D. L., *Proc. Roy. Soc. (London)*, A 249:147 (1958).
10. Bird, G. R., *J. Chem. Phys.* 28:1155 (1958).
11. Bird, G. R., and Blout, E. R., *J. Am. Chem. Soc.* 81:2499 (1959).
12. Bird, G. R., Parrish, M., and Blout, E. R., *Rev. Sci. Instr.* 29:305 (1958).
13. Bird, G. R., and Shurcliff, W. A., *J. Opt. Soc. Am.* 49:235 (1959).
14. Brecher, C., and Halford, R. S., *J. Chem. Phys.* 35:1109 (1961).
15. Brown, R. G., *J. Chem. Phys.* 38:221 (1963).
16. Bunn, C. W., *Trans. Faraday Soc.* 40:23 (1944).
17. Bunn, C. W., Chemical Crystallography, Oxford Univ. Press, (1945), p. 299.
18. Califano, S., *J. Chem. Phys.* 36:903 (1962).
19. Califano, S., *Mol. Phys.* 5:601 (1962).
20. Charney, E., *J. Opt. Soc. Am.* 45:980 (1955).
21. Cole, A. R. H., and Jones, N. R., *J. Opt. Soc. Am.* 42:348 (1952).
22. Crick, F. H. C., and Watson, J. D., *Proc. Roy. Soc. (London)*, A 223:80 (1954).
23. Daasch, L. W., *J. Mol. Spectroscopy* 8:86 (1962).

24. Delange, J., Robertson, J., and Woodward, I., *Proc. Roy. Soc. (London)*, A 171: 398 (1939).
25. Donohue, J., and Marsh, R. E., *Acta Cryst.* 15:941 (1962).
26. Elliott, A., Ambrose, E. F., and Temple, R., *J. Opt. Soc. Am.* 38:212 (1948).
27. Feughelman, M., Langridge, R., Seeds, W. E., Stokes, A. R., Wilson, H. R., Hooper, C. W., Wilkins, M. H. F., Barclay, R. K., and Hamilton, L. D., *Nature* 175:834 (1955).
28. Fraser, R. D. B., *J. Chem. Phys.* 21:1511 (1953).
29. Fraser, R. D. B., *J. Chem. Phys.* 24:89 (1956).
30. Fraser, R. D. B., *J. Chem. Phys.* 28:1113 (1958).
31. Fraser, R. D. B., and Price, W. C., *Nature* 170:490 (1952).
32. Halford, R. S., *J. Chem. Phys.* 14:8 (1946).
33. Harrick, N. J., *J. Opt. Soc. Am.* 49:376–79 (1959).
34. Hiebert, G. L., and Hornig, D. F., *J. Chem. Phys.* 20:918 (1952).
35. Hornig, D. F., *J. Chem. Phys.* 16:1063 (1948).
36. Krimm, S., "Infrared Spectra of High Polymers," in: Fortschr. Hochpolym. Forsch. 2:51 (1960).
37. Krimm, S., Liang, C. Y., and Sutherland, G. B. B. M., *J. Chem. Phys.* 25:549 (1956).
38. Krimm, S., and Schor, R., *J. Chem. Phys.* 24:922 (1956).
39. Liang, C. Y., Krimm, S., and Sutherland, G. B. B. M., *J. Chem. Phys.* 25:543 (1956).
40. Liang, C. Y., and Krimm, S., *J. Chem. Phys.* 25:563 (1956).
41. Lin, T. P., and Koenig, J. L., *J. Mol. Spectroscopy* 9:228 (1962).
42. Meyer, K. H., and Misch, L., *Helv. Chem. Acta* 20:232 (1937).
43. Migazawa, T., Fukushima, K., and Ideguchi, Y., *J. Chem. Phys.* 37:2764 (1962).
44. Mitra, S. S., "Vibration Spectra of Solids," in: Solid State Physics: Advances in Research and Applications, Vol. 13, Academic Press, New York (1962).
45. Newman, R., and Badger, R. M., *J. Chem. Phys.* 19:1147 (1951).
46. Newman, R., and Halford, R. S., *Rev. Sci. Instr.* 19:270 (1948).
47. Newman, R., and Halford, R. S., *J. Chem. Phys.* 18:1276 (1950).
48. Nielsen, J. R., and Woollett, A. H., *J. Chem. Phys.* 26:1391 (1957).
49. Novak, A., and Whalley, E., *Trans. Faraday Soc.* 55:1484 (1959).
50. Novak, A., and Whalley, E., *Trans. Faraday Soc.* 55:1490 (1959).
51. Pimentel, G. C., and McClellan, A. L., *J. Chem. Phys.* 20:270 (1952).
52. Pimentel, G. C., McClellan, A. L., Person, W. B., and Schnapp, O., *J. Chem. Phys.* 23:234 (1955).
53. Rupprecht, G., Ginsberg, D. M., and Leslie, J. D., *J. Opt. Soc. Am.* 52:665 (1962).
54. Sandeman, I., *Proc. Roy. Soc. (London)*, A 232:105 (1955).
55. Sutherland, G. B. B. M., and Tsuboi, M., *Proc. Roy. Soc. (London)*, A 239:446 (1957).
56. Tasumi, M., Shimanouchi, T., and Miyazawa, T., *J. Mol. Spectroscopy* 9:261 (1962).
57. Theilacker, W., *Z. Krist.* 90:51 (1934).
58. Tsuboi, M., *J. Polymer Sci.* 25:159 (1957).
59. Vedder, W., and Hornig, D. F., "Infrared Spectra of Crystals," in: Advances in Spectroscopy, Vol. II, Interscience Publishers, Inc., New York, (1961).

60. Waldron, R. D., and Badger, R. M., *J. Chem. Phys.* 18:566 (1950).
61. Winston, H., and Halford, R. S., *J. Chem. Phys.* 17:607 (1949).
62. Wood, D. L., and Mitra, S. S., *J. Opt. Soc. Am.* 48:537 (1958).

Infrared Spectra of Crystals

Shashanka S. Mitra

IIT Research Institute
Chicago, Illinois

Peter J. Gielisse*

Air Force Cambridge Research Laboratories
Bedford, Massachusetts

INTRODUCTION

The infrared spectrum of a solid may arise from a number of causes. We shall, however, limit our discussion here to the infrared absorption and reflection in a crystal due to the interaction of the incident electromagnetic field with its vibrational modes. Such topics as infrared spectra due to the transitions among the low-lying electronic levels of a solid, spectra due to impurities or free carriers in a semiconductor, the absorption of radiation due to the excitation of a local mode around an impurity center in a solid, and the various magneto-optic effects in semiconductors observed usually in the far infrared will not be included in the present article.

In recent years a number of reviews [1, 2, 3] have been published on the vibrational spectra of solids. However, their emphasis has been laid chiefly on the long-wavelength vibrations of the internal modes of molecular crystals. The purpose of this article is to present a unified account of the optically active vibrational transitions in solids, molecular as well as ionic, semi-ionic, and covalent crystals, and to point out the role of the $\mathbf{k} \neq 0$ ($\mathbf{k} = 2\pi/\lambda$) modes in

*Present address, General Electric Co., Metallurgical Products Dept., Detroit, Mich.

the interaction of more than one vibrational mode with the incident photon, resulting in the combination and overtone bands. We shall provide a comprehensive review of the multiphonon infrared absorption in the II-VI and III-V semiconducting crystals, which has aroused considerable interest in recent years.

No attempt will be made to describe the experimental methods of obtaining infrared transmission and reflection spectra of single-crystal specimens. It is assumed that the reader is familiar with elements of crystallography, and with the application of group theory to molecular vibration problems. These factors excluded, this paper attempts to present a complete and self-contained account. Although no special effort has been made to present a comprehensive account of the Raman spectra of solids, frequent reference to Raman measurements is made because of its complementary nature with respect to the infrared data.

NORMAL VIBRATIONS IN A CRYSTAL

A crystal may be regarded as a mechanical system of nN particles, where n is the number of particles per unit cell and N is the number of unit cells contained in the whole crystal. Such a crystal will have $3nN$ degrees freedom, of which $3nN - 3$ are the linearly independent normal modes of oscillation of the crystal and three are pure translations. The very large number ($\sim 10^{24}$) of modes belonging to a macroscopic piece of a crystal necessitates the description of the frequency spectrum in terms of a frequency distribution function.

The frequency spectrum of the nuclear motions in a solid can be determined by constructing the classical equations of motion for the lattice points and obtaining the solutions of the normal modes as plane waves. The vibration frequencies occur as the $3n$ roots of the secular equation involving the wave vector $\mathbf{k}$ ($= 2\pi/\lambda$). which may take N values.

Three of these roots approach zero as the wave vector tends to zero, and designate the acoustic branches. The remaining $3n - 3$ branches are termed optical branches, and approach finite limits as the wave vector vanishes. They constitute the fundamental vibration spectrum of the crystal. The frequency distribution $g(\nu)$ in an individual branch is obtained by calculating the number of roots in each small interval of frequency. As an elementary but illustrative example of the foregoing observations, a one-dimensional crystal consisting of two dissimilar atoms is considered.

The Linear Diatomic Chain

Let the unit cell consist of two particles, of mass M and m (as shown in Fig. 1), located at the even- and odd-numbered lattice points $2n$ and $2n + 1$, respectively. Let a be the nearest-neighbor distance. Only nearest-neighbor interactions are assumed, which can be represented by a single force constant f. The displacements u_{2n} and u_{2n+1} of the even and odd particles are given by the following equations of motion:

$$M\ddot{u}_{2n} = f(u_{2n+1} + u_{2n-1} - 2u_{2n})$$
$$m\ddot{u}_{2n+1} = f(u_{2n+2} + u_{2n} - 2u_{2n+1}) \tag{1}$$

Assuming solutions of the form

$$u_{2n} = y_1 \exp i(2\pi\nu t + 2nka)$$
$$u_{2n+1} = y_2 \exp i[2\pi\nu t + (2n + 1) ka] \tag{2}$$

and substituting them in equation (1), one obtains two equations for the amplitudes y_1 and y_2. The compatibility condition for these equations is

$$\begin{vmatrix} 2f - 4\pi^2\nu^2 M & -2f\cos ka \\ -2f\cos ka & 2f - 4\pi^2\nu^2 m \end{vmatrix} = 0 \tag{3}$$

which gives

$$\nu^2 = \frac{1}{4\pi^2} \frac{f}{\mu} \pm \left[\frac{f^2}{\mu^2} - \frac{4f^2 \sin^2 ka}{Mm}\right]^{1/2} = 0 \tag{4}$$

where μ is the reduced mass per unit cell. This relation between ν and the wave vector $\mathbf{k}$ is known as the dispersion relation. The finite length of the lattice $2Na$ restricts the possible values of k in the range $-\pi/2a \le k \le \pi/2a$. The region between these limits of k is termed the first Brillouin zone.

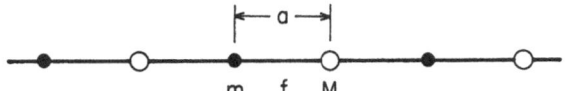

$$m \quad f \quad M$$

Fig. 1. Linear diatomic chain.

Equation (4) has two sets of solutions, depending on the positive or the negative signs, which are known as the optical and acoustic branches. In the long-wave limit (small k) the two roots are

$$\nu = \frac{1}{2\pi} \left(\frac{2f}{\mu} \right)^{1/2} \qquad \text{(optical)} \qquad (5)$$

$$\nu = \frac{1}{2\pi} \left(\frac{2f}{M + m} \right)^{1/2} ka \qquad \text{(acoustic)} \qquad (6)$$

As $k \to 0$, the frequency of the acoustic branch tends to zero, as expected, since it then represents the motion of the crystal as a whole. It can be shown that the frequency distribution function $g(\nu)$, defined such that $g(\nu)d\nu$ gives the number of frequencies between ν and $\nu + d\nu$, has maxima at $(1/2\pi)(2f/m)^{1/2}$ and $(1/2\pi)(2f/M)^{1/2}$ corresponding to the boundary of the Brillouin zone, $k = \pi/2a$. The dispersion relations for the two branches are shown in Fig. 2. The frequency distribtuion for the diatomic linear lattice is given in Fig. 3.

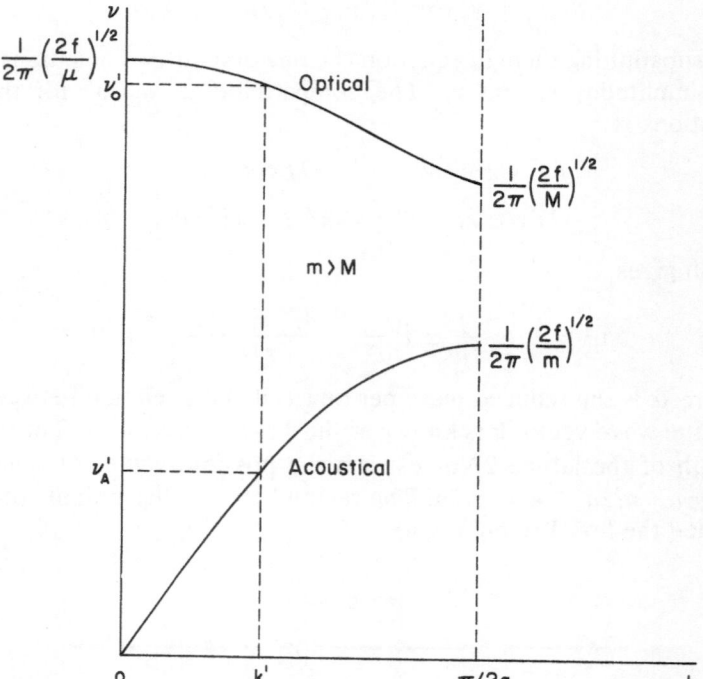

Fig. 2. The dispersion relation for a linear diatomic chain.

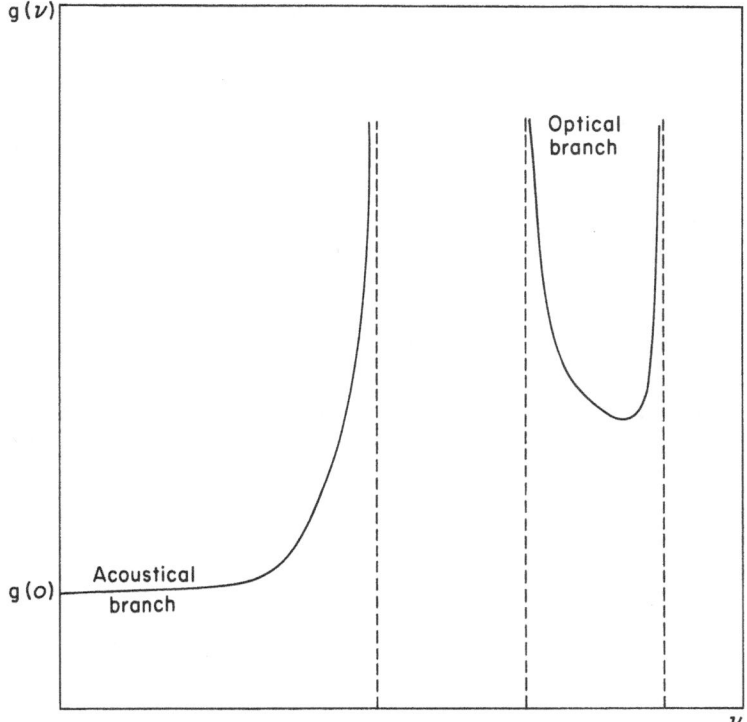

g(ν)

g(o)

Optical branch

Acoustical branch

ν

Fig. 3. The frequency distribution function for a linear diatomic chain.

When electromagnetic radiation interacts with a crystal lattice, a photon may be absorbed, provided that the wave vector and the energy are conserved. For infrared radiation of wavelength $\sim 50~\mu$, the wave vector $k = 2\pi/\lambda \approx 10^3$ cm^{-1}. This is extremely small compared with the wave vector corresponding to the edge of the Brillouin zone, $\pi/2a \simeq 10^8$ cm^{-1}. Thus, one may consider the wave vector associated with the electromagnetic field in the infrared region to be essentially zero. The infrared spectrum for a diatomic linear lattice, in the long-wave limit ($k \simeq 0$) corresponding to the center of the Brillouin zone, will then consist of a single line at a frequency equal to $1/2\pi~(2f/\mu)^{1/2}$, provided it is associated with a changing transition moment. This fundamental infrared absorption originates from the relative motion of the M lattice as a whole against the m lattice. For a homopolar diatomic linear lattice this fundamental mode is forbidden in the infrared, and is active only in Raman scattering.

Vibration of Three-Dimensional Lattices

Analogous equations of motion can be constructed for the two- and three-dimensional lattices also. Such equations were first set up by Born and von Kármán [4]. The solutions obtained are in terms of plane waves of the form $\exp i(2\pi\nu t - \mathbf{k} \cdot \mathbf{r}_j)$. For a monatomic lattice there are N independent values of $\mathbf{k}$, each of which is associated with three different modes having different and orthogonal polarization directions. Thus, there are $3N$ independent lattice modes. The energy of each mode is quantized and the term *phonon* is used to describe a quantized lattice vibration. For a simple Bravais lattice (one particle per unit cell) such as sodium or copper, all the branches have zero frequencies in the long-wave limit. These, therefore, do not have any optical vibration spectra. In a lattice with basis, i.e., in which the entire crystal is obtained by the primitive translations of a unit cell containing more than one particle, the optical branches are separated from the acoustic ones, and the former have nonvanishing frequencies at $\mathbf{k} = 0$. These constitute the optical spectrum.

To sum up, a crystal composed of N primitive cells with n atoms per cell has $3Nn$ normal modes of vibration. These $3Nn$ modes are distributed on $3n$ branches, $3n - 3$ of which are optical and 3 acoustic branches. In each branch the mode frequency depends upon the wave vector, which can assume all values in and on the Brillouin zone. However, all modes of vibration, other than those in which equivalent atoms move identically in phase, are forbidden as fundamentals both in infrared absorption and in Raman scattering. This is due to the fact that the wave vector associated with a photon is essentially zero compared with that of most of the phonons of the same energy, except for the ones near the center of the Brillouin zone. The three acoustic modes have vanishing frequencies at $\mathbf{k} = 0$. The $3n - 3$ optical-branch frequencies corresponding to $\mathbf{k} = 0$ constitute the fundamental modes of vibration, whose distribution, symmetry classification, and optical activity can be enumerated only from a consideration of the unit cell instead of the entire crystal.

For example, in rock salt the number of zone center ($\mathbf{k} = 0$) fundamentals is equal to $3 \times 2 - 3 = 3$. Of the three fundamentals, one corresponds to the longitudinal optical mode and the other two to the doubly degenerate transverse optical modes. The electromagnetic waves, because they are transverse, cannot interact with longitudinal phonons in an infinite crystal, and thus the longitudinal optical-mode frequency is inactive in the infrared. The transverse optical-mode frequency, on the other hand, is optically active, and is observed as a resonant frequency by measuring the transmittance of thin films for infrared radiation, or as the reststrahlen band in the reflection spectrum. For homopolar crystals such as Ge or Si, the transverse optical mode is only active in Raman scattering.

Symmetry and Unit Cell Modes

In this section attention is confined to the modes of zero wave vector, i.e., the ones corresponding to the center of the Brillouin zone. It was mentioned earlier that these are the only modes that may be optically active as fundamentals. They are $3n - 3$ in number, where n is the number of particles per Bravais unit cell. The equivalent particles in the crystal lattice move in identical phase for these modes of oscillation. In other words, all the unit cells of the lattice are in the same phase throughout the periods of execution of the fundamental optical modes. The symmetry classification of the $k = 0$ modes of vibration can therefore be accomplished by taking into consideration the factor group of the crystal, instead of the entire space group. The factor group represents the group formed by the symmetry elements contained in the smallest unit cell (Bravais cell). A factor group is always isomorphous with one of the 32 crystallographic point groups. The $3n - 3$ long-wavelength fundamental modes of a crystal are also known as the factor group fundamentals.

In the case of crystals consisting of a number of structural units, it may be assumed that the forces holding together the atoms within a group are much stronger than those keeping the various groups together. For example, in a single crystal of naphthalene the carbon and hydrogen atoms in a naphthalene molecule are held together by chemcial bonds, whereas the individual molecules in the crystal are held chiefly by van-der-Waals-type binding. The former is about two orders of magnitude larger than the latter. The modes arising chiefly from the oscillation of atoms within a group are termed *internal* modes. The motions of the groups relative to one another give rise to the *external* modes, which are also known as *lattice* modes. Since the forces among the groups are usually weaker than those among the atoms within a group, the external vibrations are expected to occur at lower frequencies than the internal ones. The external modes can be further subdivided into translatory and rotatory types, which in the limit of vanishing forces among the groups correspond to pure translations and rotations, respectively. The rotatory lattice modes are often referred to as *librational* modes, and may be associated with any polyatomic group in a crystal. In a crystal like rock salt, the only possible vibrational modes are lattice modes. Because of the lack of any molecular group in such a crystal, no internal mode occurs.

The determination of selection rules and the symmetry classification of the unit cell fundamentals of a crystal can be accomplished by two methods. Bhagavantam and Venkatarayudu [5] consider the atoms in the unit cell as a large molecule, and the usual method [6] for deriving the optical activities and the symmetry classification of the normal modes of a molecule is applied. Halford [7], on the other hand, suggested use of the site approximation in which the local symmetry of a molecular group in the unit cell is considered. Hornig [8] has shown that proceeding from the site approximation one may obtain the same results as Bhagavantam and Venkatarayudu. Winston and Halford [9] have derived both methods by considering the motions of a crystal segment composed of an arbitrary number of unit cells and subject to the Born-von Kármán boundary conditions. The factor group analysis of Bhagavantam and Venkatarayudu is most suitable for classifying the lattice vibrations. Mitra [10] has indicated an extension of the method which is suitable for crystals containing linear polyatomic groups. A brief summary of the method is presented here.

The procedure consists of writing down the character table and irreducible representations of the isomorphous point group corresponding to the space group of the crystal. N_k, the number of times a particular irreducible representation Γ_k is contained in another representation Γ, is given by

$$N_k = \frac{1}{N} \sum_j h_j \chi_k(R) \chi_j'(R) \tag{7}$$

where N is the order of the group and h_j the number of group operations contained in the jth subgroup. $\chi_k(R)$ and $\chi_j'(R)$ are, respectively, the characters of the group operation R in the representations

Γ_k and Γ. The normal modes are classified by making specific selections of the representation Γ and its appropriate characters $\chi'_j(R)$. One needs to consider the characters for the electric dipole moment vector (T) and the polarizability or the symmetric tensor (α) representations, respectively, in order to obtain the infrared- and Raman-active fundamentals. The selection rules can be summed up as

$$\frac{1}{N} \sum_j h_j \chi_k(R) \chi'_j(T) \quad \left\{ \begin{array}{l} = 0 \text{ IR forbidden} \\ \neq 0 \text{ IR permitted} \end{array} \right. \tag{8}$$

and

$$\frac{1}{N} \sum_j h_j \chi_k(R) \chi'_j(\alpha) \quad \left\{ \begin{array}{l} = 0 \text{ Raman forbidden} \\ \neq 0 \text{ Raman permitted} \end{array} \right. \tag{9}$$

The group characters $\chi'_j(R)$ for the various representations are given in Table I. Two special cases of selection rules applicable to

TABLE I
Group Characters χ'_j (R) for Various Representations

Representation	Group character[a]
All unit cell modes (3n Cartesian coordinates)	$\chi'_j(n_i) = \omega_R(\pm 1 + 2 \cos \phi_R)$
Acoustic modes (Dipole moment vector)	$\chi'_j(T) = \pm 1 + 2 \cos \phi_R$
Translatory lattice modes	$\chi'_j(T') = [\omega_R(s) - 1] (\pm 1 + 2 \cos \phi_R)$
Rotatory lattice modes	$\chi'_j(R') = [\omega_R(s - p)] \chi'_j (P)$
Symmetric tensor	$\chi'(\alpha) = (\pm 1 + 2 \cos \phi_R) 2 \cos \phi_R$

[a] ω_R is the number of atoms invariant under the symmetry operation R; $\omega_R(s)$ is the number of structural groups remaining invariant under symmetry operation R; $\omega_R(s - p)$ is the number of polyatomic groups remaining invariant under an operation R (p is the number of monatomic groups); ϕ_R is the angle of rotation corresponding to the symmetry operation R (plus and minus signs stand, respectively, for proper and improper rotations); $\chi'_j (P)$ is equal to $(1 \pm 2 \cos \phi_R)$ for nonlinear polyatomic groups, it is equal to $\pm 2 \cos \phi_R$ for operations $C (\phi_R)$ and $S (\phi_R)$ in a linear polyatomic group, and it is equal to 0 for operations $C_2 (\theta)$ and σ_v in a linear polyatomic group.

the factor group fundamentals are worth remembering. First, a totally symmetric mode is always Raman active. Second, in a crystal with a point of inversion, a normal mode symmetric with respect to the center of symmetry is Raman active and forbidden in the infrared, whereas an antisymmetric mode with respect to the center is infrared active but forbidden in Raman scattering.

Normal Modes of Oscillation in Gypsum

Factor Group Analysis. The foregoing considerations will now be used in the elucidation of the factor group fundamentals of gypsum, $CaSO_4 \cdot 2H_2O$. The crystal belongs to the space group $C_{2h}^6(C2/c)$ with two molecules per Bravais unit cell. Gypsum is thus composed of three types of structural groups: monatomic ions, Ca^{2+}, polyatomic ions, SO_4^{2-}, and H_2O molecules. The structure is shown in Figs. 4 and 5. Exclusive of the long-wavelength acoustic modes, there are a total of 69 factor group fundamentals. The internal modes belong to the sulfate ions and the water molecules only. They are expected to occur in the crystal at nearly the same frequencies as for the isolated systems SO_4^{2-} and H_2O. The lattice modes are caused by the motions of the structural groups (eight in number in the case of gypsum) relative to each other. The

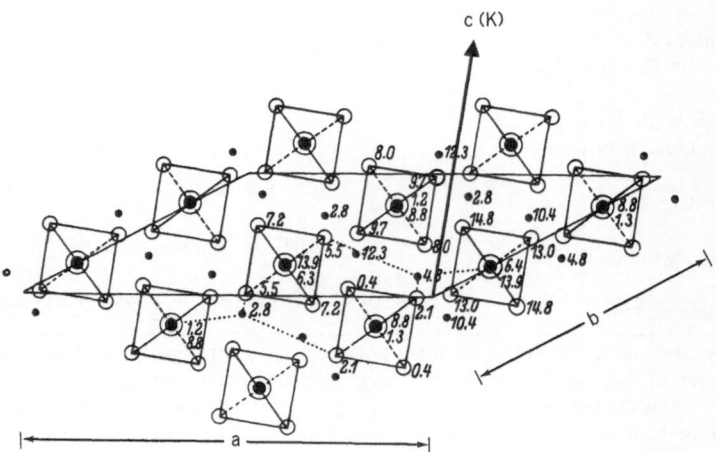

Fig. 4. Projection of the crystal structure of gypsum on the (010) plane. Heights are in 10^{-8} cm. Squares indicate SO_4 tetrahedra. The upper of the two numbers beside the symbol ⊙ gives the height of the S. [Reprinted by permission from W. A. Wooster, *Z. Krist.* 94:375 (1936).]

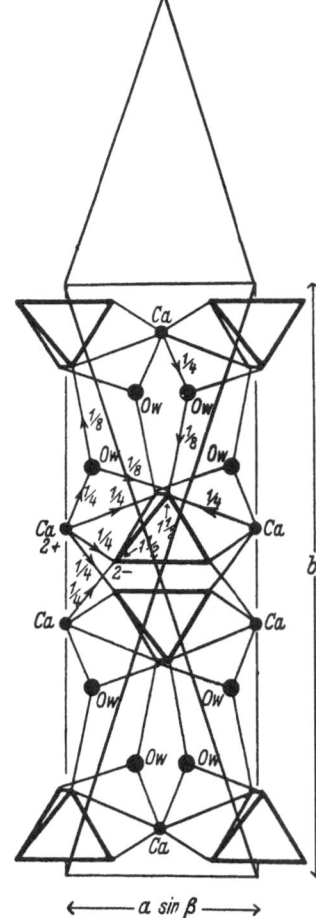

Fig. 5. Projection of the gypsum structure with the Wooster c axis normal to the paper. Numbers indicate the disturbance of valence bonds. [Reprinted by permission from W. A. Wooster, *Z. Krist.* 94:375 (1936).] The diamond shaped outline (drawn over Wooster's diagram) shows the Bravais unit cell.

classification of the various modes of vibrations has been carried out by means of the factor group analysis shown in Table II. This readily gives the symmetry species and the activities of the external modes. The number of internal modes belonging to the SO_4^{2-} and H_2O groups is obtained by subtracting the number of external modes from the total number of modes belonging to any particular irreducible representation. In order to ascertain the number of internal modes belonging to SO_4^{2-} and H_2O separately, an artifice due to Hass and Sutherland [11] is most useful. The lattice is first considered to be composed only of $CaSO_4$ to obtain the sulfate

TABLE II
Character Table and Distribution of Unit Cell Modes for Gypsum

C_{2h}^6	E	C_2	i	σ_h	n_i	T	T'	R'	n_i' SO$_4^{2-}$	H$_2$O	IR	Raman
A_g	1	1	1	1	17	0	5	4	5	3	f	$p\,(\alpha_{xx},\,\alpha_{yy},\,\alpha_{zz},\,\alpha_{xz})$
A_u	1	1	-1	-1	17	1	4	4	5	3	$p\,M_y$	f
B_g	1	-1	1	-1	19	0	7	5	4	3	f	$p\,(\alpha_{xy},\,\alpha_{yz})$
B_u	1	-1	-1	1	19	2	5	5	4	3	$p\,M_x,\,M_y$	f
ϕ_R	0°	180°	180°	0°								
ω_R	24	4	0	0								
$\omega_R(s)$	8	4	0	0								
$\omega_R(s-p)$	6	2	0	0								
h_j	1	1	1	1								
$\chi_j'(n_i)$	72	-4	0	0								
$\chi_j'(T)$	3	-1	-3	1								
$\chi_j'(T')$	21	-3	3	-1								
$\chi_j'(R')$	18	-2	0	0								
$\chi_j'(\alpha)$	6	2	6	2								

fundamentals. Next the lattice is regarded as consisting only of water molecules, ignoring the $CaSO_4$. This procedure is justified since the number of internal modes of each kind of molecule depends only on the number of that kind of molecule in the unit cell.

Internal Modes of the Sulfate Ions. The internal vibrations of the sulfate ions in gypsum will now be discussed by the site group method of Halford [7]. The free sulfate ion has a tetrahedral symmetry. The symmetry species and the designation of the normal modes of vibration of an XY_4 molecule belonging to the T_d point group are given in Table III. There are four distinct modes for the free molecule. The nondegenerate $\nu_1(A_1)$ and the doubly degenerate $\nu_2(E)$ modes are Raman active and are forbidden in the infrared. The two triply degenerate modes $\nu_3(F_2)$ and $\nu_4(F_2)$ are active both in Raman and infrared. The normal modes of vibration are shown in Fig. 6.

In the gypsum crystal the C_2 crystal axis coincides with a C_2 axis of each SO_4^{2-} group. The motions of the sulfate ions in the crystal thus have to be classified as those of a molecule with C_2 symmetry only. However, the intermolecular forces giving rise to the reduction in symmetry in the crystal are weak compared with the intramolecular forces within an SO_4^{2-} group. The internal modes of the ions may therefore be considered not very different from those of a tetrahedral molecule under a small perturbation having the symmetry of the lattice, causing a breakdown of the degeneracies. The symmetry species of the C_2 point group are given in Table IV. The fundamental modes of the free molecule are distributed among the two irreducible representations of the C_2 point group as shown in the correlation chart of Fig. 7. Since the $\nu_1(A_1)$ and $\nu_2(E)$ fundamentals of the free molecule are completely symmetric with respect to the twofold rotation operation, they correlate with the A species of the C_2 point group. They thus become permitted fundamentals in the infrared, with the transition moment parallel to the $C_2(y)$ axis. If the coupling forces are weak, these modes will be only weakly active in the infrared and may not appear at all in the case of vanishing coupling. The triply degenerate F_2 modes, ν_3 and ν_4 of the free molecule, now split into three single frequencies each. For each set, one of the modes is symmetric (A) and the other two are antisymmetric with respect to the twofold axis. Correspondingly, the former transition moment lies along the $C_2(y)$ axis and the latter lies perpendicular (x and z) to the C_2 axis.

TABLE III

Character Table and Distribution of Normal Modes of an XY$_4$ Molecule in the T_d Point Group

T_d	E	$8C_3$	$6\sigma_d$	$6S_4$	$3S_4^2 = 3C_2$	n	Designation	Activity IR	Activity Raman
A_1	1	1	1	1	1	1	ν_1	f	$p\,(\alpha_{xx},\alpha_{yy},\alpha_{zz})$
A_2	1	1	-1	-1	1	0		f	f
E	2	-1	0	0	2	1	ν_2	f	$p\,(\alpha_{xx},\alpha_{yy},\alpha_{zz})$
F_1	3	0	-1	1	-1	0		f	f
F_2	3	0	1	-1	-1	2	ν_3,ν_4	$p\ Tx;Ty;\ Tz$	$p\,(\alpha_{xy},\alpha_{yz},\alpha_{zx})$

TABLE IV

Character Table and Distribution of Normal Modes of an XY$_4$ Molecule in the C_2 Point Group

C_2	E	$C_2(y)$	n	Activity IR	Activity Raman
A	1	1	5	$p\ T_y$	$p\,(\alpha_{xx},\alpha_{yy},\alpha_{zz},\alpha_{zx})$
B	1	-1	4	$p\ T_x;T_z$	$p\,(\alpha_{xy},\alpha_{yz})$

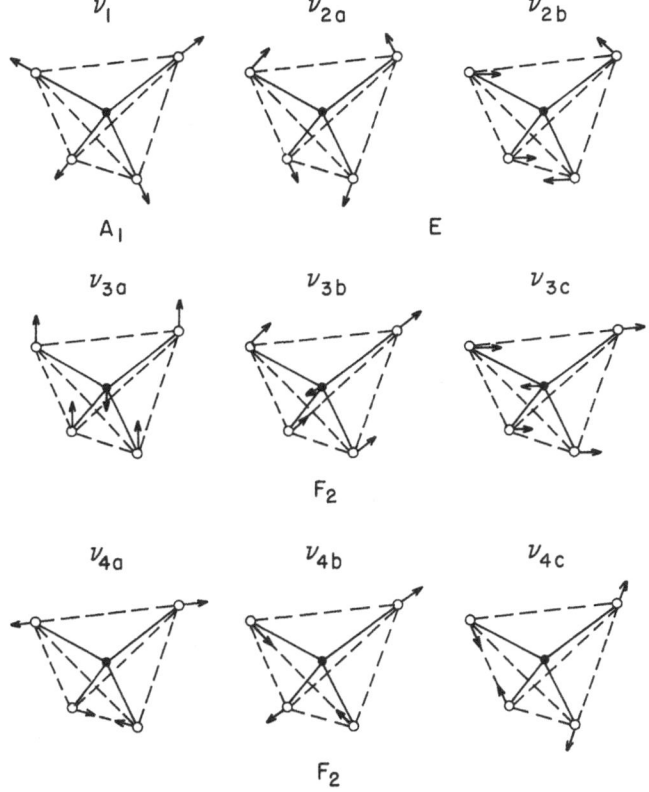

Fig. 6. Normal vibrations of a tetrahedral XY_4 molecule.

In the above considerations only one SO_4^{2-} ion was subjected to the perturbation having the symmetry of the lattice. It is, however, necessary to investigate the motions of all the molecules in the unit cell. In the present case, there are two SO_4^{2-} ions per Bravais cell, each capable of undergoing nine fundamental modes of oscillation. Each mode therefore is further split into two, depending on whether it is symmetric (g) of antisymmetric (u) to the point of inversion in the unit cell. This is shown in the last column of the correlation diagram, which is identical with the distribution of the internal modes (n') of the SO_4^{2-} ions in Table II. Thus, although there are only four normal modes possible in the free tetrahedral SO_4^{2-} ion, there may be as many as eighteen in the crystal corresponding to the motions of all unit cells in phase ($\mathbf{k} = 0$).

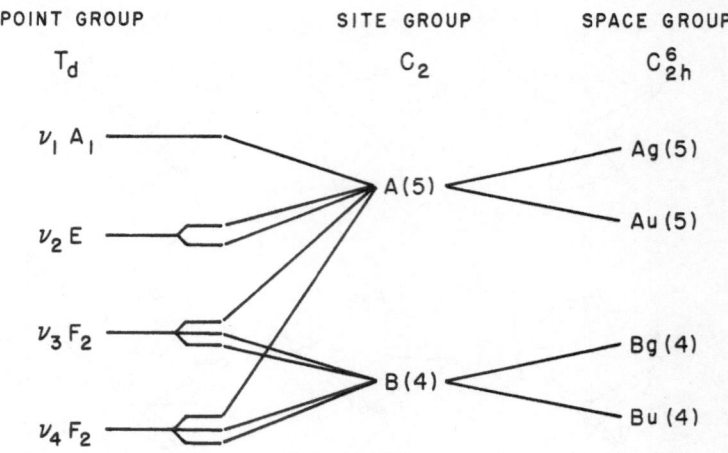

Fig. 7. Correlation chart for the SO$_4$ vibrations.

Internal Modes of the Water Molecule. The three fundamental modes of vibration due to an isolated water molecule are shown in Fig. 8. All of them are active both in the infrared and the Raman spectrum. The modes ν_1 and ν_2 are symmetric with respect to the molecular symmetry axis with the transition moments parallel to this direction, whereas the ν_3 mode is antisymmetric with transition moment normal to the molecular symmetry axis.

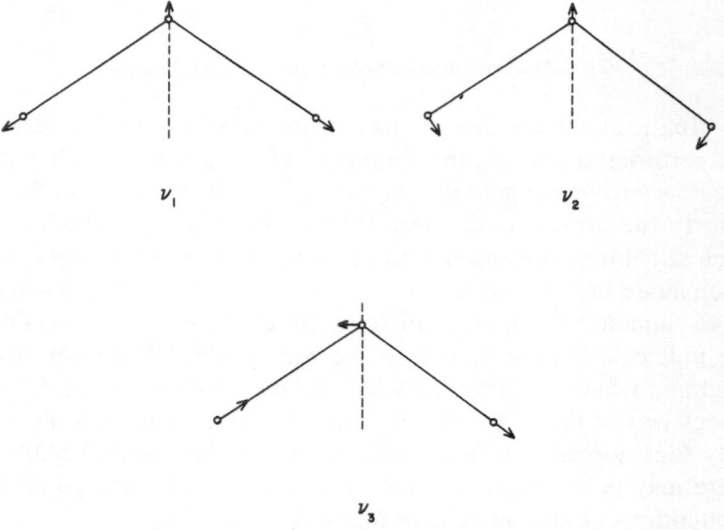

Fig. 8. Normal modes of vibration of a water molecule.

There are four molecules of water of crystallization per unit cell of gypsum. The crystal internal modes may now be constructed by the superposition of the molecular modes of the four water molecules in the unit cell in such a way that they satisfy the symmetry requirements of the crystal. In the absence of any interaction among the molecules, each molecular mode becomes fourfold degenerate in the crystal. Introduction of a small amount of coupling leads to a breakdown of degeneracy. Each molecular fundamental splits into four distinct crystal modes, belonging one each to the four symmetry species of the C_{2h} factor group.

The four crystal modes, for example, those corresponding to the ν_1 frequency of H_2O, should occur at slightly different frequencies. Since the extent of the coupling among the water molecules is not known, the crystal modes may be regarded as arising from the limiting motions of an array of molecules oriented in the lattice positions but with vanishing interactions among them. Such a model, known as the *oriented-gas model*, was first proposed by Ambrose, Elliot, and Temple [12], and Pimentel and McClellan [13]. Though this model is incapable of predicting the extent of splitting of the molecular modes in the crystal, it is useful in deducing the relative intensities of the various crystal fundamentals. Let the angle θ denote the orientation of the water symmetry axis with respect to the b axis of the crystal as indicated in Fig. 9. The ratio of the transition moments of the A_u and B_u bands arising from the ν_1 and ν_2 molecular modes will be given by cot θ. Hence, the ratio of the extinction coefficients of the $A_u(\|)$ to $B_u(\perp)$ bands known as the *dichroic ratio** is given by cot² θ. The angle θ in gypsum, as determined from x-ray diffraction data, is 52°, predicting a dichroic ratio of 0.613 for the A_u and B_u crystal modes corresponding to the symmetric (ν_1 and ν_2) molecular vibrations. A similar analysis can be carried out for the antisymmetric ν_3 molecular vibration as well. There, the pertinent angle is that between the line joining the hydrogen nuclei in a water molecule and the b axis, which is 40° in gypsum. Hence, the predicted dichroic ratio for the ν_3 water fundamental in gypsum is 1.40.

The Effect of Crystal Field on the Internal Modes. When compared with the spectrum due to an isolated molecule the infrared spectrum of a solid has the following distinctive characteristics:

---

*For a detailed discussion of polarized infrared spectroscopy see Ref. 1 pp. 22 – 29, and Ref. 14.

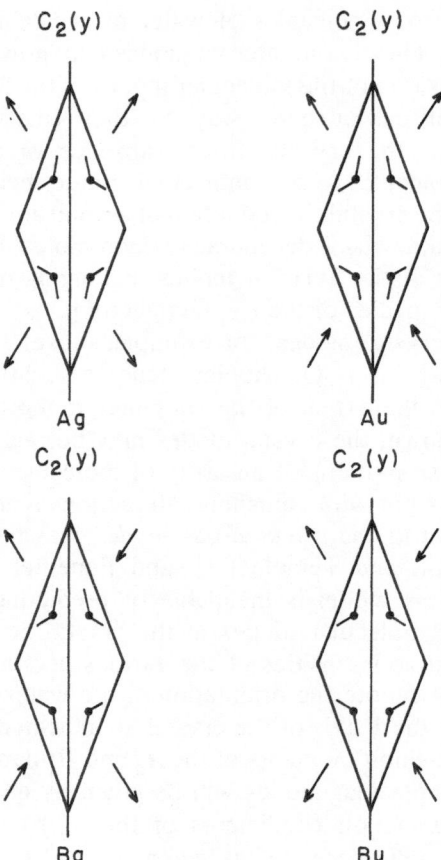

Fig. 9. ν_1 and ν_2 modes of water molecules in gypsum.

(i) appearance of entirely new brands in the low-frequency region (< 800 cm^{-1}) due to the external modes, (ii) changes in vibration frequencies, occasionally accompained by changes of intensities, and (iii) splitting of single bands into multiplets. In the present section, it is intended to furnish a qualitative explanation of the last two features.

The concept of an isolated molecule may nearly be achieved in a gas at low pressure, and to a certain extent in a dilute solution with an inert solvent. The interactions with other molecules may be entirely neglected in such a situation, and the observed spectrum may be explained from the consideration of the potential energy for the internal coordinates of a single molecule. In the liquid and the solid state, the interactions with the environment in general, and the immediate neighbors in particular are expected to have perturbing effects, causing changes in the spectrum. Thus, by investigating the modifications occurring in the spectrum while going from the vapor to the liquid and solid phases, it is possible in principle to obtain some information regarding the nature of the intermolecular forces in the condensed states.

The vibrational potential energy of a molecular crystal, in general, may be represented [15] by

$$V = \sum_{n} V_n + \sum_{n} \sum_{k} V_{nk} + V_l + V_{ln} \qquad (10)$$

where V_n is the potential energy due to all the internal coordinates of the nth molecular group in the crystal; $\sum_{n} V_n$ is the sum of the internal potential energy for all the molecules contained in the crystal; V_{nk} are the potential energy cross terms between the internal coordinates of the nth and the kth molecules; $\sum_{n} \sum_{k}$ is the sum over all pairs of molecules; V_l is the potential arising from the external coordinates, i.e., due to the displacements of the centers of gravity and changes in orientations of the various structural units in the crystal; and V_{ln} are the cross terms involving the internal and the lattice coordinates.

In the oriented-gas approximation only the first term need be considered. However, the potential energy for a molecule in the crystal will be different from that for the corresponding gas molecule due to the so-called *static field effect*. This results in the usual difference in frequencies noted between the corresponding gas and condensed phase modes, and includes the influence of the surrounding lattice in its equilibrium configuration on the molecule in question. V_n thus incorporates the site symmetry of this molecule. If the site group has a lower symmetry than the molecular point group, intrinsic degeneracy, if any, of the molecular modes no longer exists in the crystal as was observed in the case of the sulfate vibrations in gypsum. Splitting will result in the presence of a nonvanishing static field effect, in addition to the shift of frequencies relative to those of the free molecule. It should be borne in mind that a comparison of spectra in the solid and gas phases may be somewhat hypothetical in nature. In many instances the isolated molecule can not be studied, as is the case with ions in a solid.

An approximate theory explaining the condensed phase frequency shifts was given by West and Edwards [16], and Bauer and Magat [17]. Considering the overall effect of the environment, they obtained the following relation:

$$\frac{\Delta\nu}{\nu_0} = C \frac{\epsilon - 1}{2\epsilon + 1} \tag{11}$$

where ν_0 is the gas phase frequency, $\Delta\nu$ the frequency shift, and ϵ the dielectric constant of the medium, and C a property of the medium determined empirically. Since only the bulk properties of the medium are included in relation (11), departures from it are very frequent [18]. Pullin [19] has made some advances by considering the influence of the medium involving the anharmonic terms in vibrational potential energy and higher-order terms in the electric moment. However, the theory is of limited applicability, because the expression for the frequency shift contains far too many undeterminable parameters. Buckingham [20] has given a quantum-mechanical theory for a diatomic molecule in a solvent. The interaction potential of the solvent – solute system is expressed as a power series of the internuclear displacement of the atoms of the solute molecule. It is introduced as a perturbation on the anharmonic oscillator Hamiltonian. An analogous theory for the solid is possible which replaces the solvent by the lattice itself. However, no complete theory has yet been advanced explaining the static field splitting satisfactorily. This is partly because the parameters involved usually far outnumber the experimentally determinable quantities.

The cross terms $\sum_n \sum_k V_{nk}$ represent the *dynamic crystal effects* arising from the intermolecular interactions. The most important of these is the so-called *correlation field splitting* associated with the site-group-to-factor-group transformation. For example, the single OH^- ion has only one internal mode due to the hydroxyl stretching vibration. However, in the brucite crystal $Mg(OH)_2$, there are two internal modes due to OH stretching which in turn results from two OH^- ions per unit cell. In fact one expects two different crystal modes to exist for each **k** value. The magnitude of the splitting depends upon the extent of interaction between one OH^- ion in a unit cell with all the other nonequivalent OH^- ions in the entire crystal. For the **k** = 0 internal modes due to the water molecules in gypsum, such a splitting was noted earlier.

The static and the dynamic field effects are usually present simultaneously. Hexter [21], for example, has examined methyl chloride, CH_3Cl, for this condition. He has predicted that a nondegenerate (A_1) internal fundamental will be split into two infrared-active frequencies by the intermolecular coupling (correlation), and a doubly degenerate (E) fundamental will be split in two by the site symmetry C_S (static field effect), in addition to a further doubling by the intermolecular coupling. However, only three of the four crystal modes of the latter type are infrared active under the space group C_{2v}. Dows [22] has observed the correlation field splitting for the nondegenerate modes ν_1, and ν_2 of CH_3Cl, and ν_2 and ν_3 of CH_3Br and CH_3I. The ν_3 of CH_3Cl shows an isotope shift of 6 cm^{-1} in the gas phase due to the great abundance of Cl^{35} and Cl^{37}. In the crystal, three lines are observed, presumably due to a superposition of the isotope effect and the correlation field shift, with the lower of the Cl^{35} components coinciding with the upper of the Cl^{37} components. The absorption bands due to the crystal modes derived from the doubly degenerate E modes are more difficult to interpret. The ν_6 of CH_3Cl is not split, while in CH_3Br and CH_3I the observed splitting amounts to 5 and 7 cm^{-1}, respectively. The site splitting is expected to be more prominent in CH_3Cl than in CH_3Br and CH_3I because of the larger dipole moment and closer packing factor in the former. The splittings, however, increase in the reverse order, indicating that the intermolecular coupling is the predominant source of splitting. Except for the carbon–halogen stretching mode, the splittings cannot be accounted for by dipole–dipole interactions alone. Dows has shown [23] that an intermolecular hydrogen–hydrogen repulsion potential satisfactorily explains the observed effects in the case of the modes involving the hydrogen motions.

INFRARED DISPERSION BY IONIC CRYSTALS

It may be recalled that in the long-wave limit the optical vibrations of a diatomic lattice correspond to the motion of one type of atoms, all in phase, relative to the other kind. In ionic crystals, such a motion is associated with strong electric moments and hence can directly interact with the electric field of proper polarization from incident electromagnetic radiation. In the vicinity of the resonance frequency, one thus expects drastic changes in the optical properties of such a crystal. In this section it is intended to discuss briefly the dispersion of infrared radiation by cubic diatomic ionic crystals with optical isotropy.

Interaction with the Radiation Field

Huang [24] has given a phenomenological theory of infrared dispersion in ionic crystals, salient features of which are described below. If $\mathbf{u}$ represent the displacement of the positive ions relative to the negative ions, a reduced displacement vector $\mathbf{w}$ may be expressed as $\mathbf{w} = \mathbf{u} \, (\mu/v_a)^{1/2}$, where μ/v_a is the reduced mass per Bravais unit cell. The macroscopic equations describing the polar motion are then

$$\ddot{\mathbf{w}} = b_{11}\mathbf{w} + b_{12}\mathbf{E} \tag{12}$$

and

$$\mathbf{P} = b_{21}\mathbf{w} + b_{22}\mathbf{E} \tag{13}$$

where $\mathbf{E}$ is the electric field and $\mathbf{P}$ the dielectric polarization defined by

$$\mathbf{E} + 4\pi\mathbf{P} = \mathbf{D} = \epsilon\mathbf{E} \tag{14}$$

The b's are constants characteristic of the solid the nature of which is to be ascertained. It can be shown [25] that $b_{12} = b_{21}$, as a consequence of the principle of conservation of energy. The linearity of the Eqs. (12) and (13) implies that anharmonicity and higher-order terms in the electric moment are neglected.

Considering the periodic solutions

$$(\mathbf{w}, \mathbf{E}, \mathbf{P}) = (\mathbf{w}_0, \mathbf{E}_0, \mathbf{P}_0) \, e^{2\pi i \nu t} \tag{15}$$

Eqs. (12) and (13) are reduced to

$$-4\pi^2\nu^2\mathbf{w} = b_{11}\mathbf{w} + b_{12}\mathbf{E} \tag{16}$$

and

$$\mathbf{P} = b_{21}\mathbf{w} + b_{22}\mathbf{E} \tag{17}$$

Elimination of **w** from Eqs. (16) and (17) yields

$$\mathbf{P} = \left\{ b_{22} + \frac{b_{12}b_{21}}{-b_{11}-4\pi^2\nu^2} \right\} \mathbf{E} \tag{18}$$

Substitution of Eq. (18) in Eq. (14) readily gives the dielectric constant ϵ in terms of the b coefficients:

$$\epsilon = 1 + 4\pi b_{22} + \frac{4\pi b_{12} b_{21}}{-b_{11}-4\pi^2\nu^2} \tag{19}$$

The similarity of Eq. (19) to the infrared dispersion formula

$$\epsilon = \epsilon_\infty + \frac{(\epsilon_0 - \epsilon_\infty)\nu_0^2}{\nu_0^2 - \nu^2} \tag{20}$$

is obvious. In Eq. (20), ν_0 is the dispersion frequency; ϵ_∞ is equal to n^2; n is the refractive index for light of $\nu \gg \nu_0$ in a nondispersive region; and ϵ_0 is the d.c. or low-frequency dielectric constant. The b coefficients may now be obtained by comparing Eq. (19) with Eq. (20):

$$b_{11} = -4\pi^2\nu_0^2 \tag{21}$$

$$b_{12} = b_{21} = (\epsilon_0 - \epsilon_\infty)^{1/2}\nu_0 \tag{22}$$

and

$$b_{22} = \frac{\epsilon_\infty - 1}{4\pi} \tag{23}$$

We have, yet however, to identify the dispersion frequency ν_0.

Since macroscopically the crystal is electrically neutral, one can apply Gauss's law

$$\nabla \cdot \mathbf{D} = \nabla \cdot (\mathbf{E} + 4\pi\mathbf{P}) = 0 \tag{24}$$

P can then be eliminated from Eq. (13), giving

$$\nabla \cdot \mathbf{E} = \frac{-4\pi b_{21}}{1 + 4\pi b_{22}} \nabla \cdot \mathbf{w} \tag{25}$$

The solenoidal and irrotational solutions of this equation correspond to the transverse and longitudinal waves, respectively, for which

$$\nabla \cdot \mathbf{w}_t = 0 \qquad \text{(solenoidal)} \tag{26}$$

and

$$\nabla \times \mathbf{w}_l = 0 \qquad \text{(irrotational)} \qquad (27)$$

where

$$\mathbf{w} = \mathbf{w}_t + \mathbf{w}_l \qquad (28)$$

Consequently the equation of motion (12) can be split into two parts

$$\ddot{\mathbf{w}}_t = b_{11}\mathbf{w}_t = -4\pi^2\nu_0^2\mathbf{w}_t \qquad (29)$$

and

$$\ddot{\mathbf{w}}_l = \left\{ b_{11} - \frac{4\pi b_{12}b_{21}}{1 + 4\pi b_{22}} \right\} \mathbf{w}_l = -\left(\frac{\epsilon_0}{\epsilon_\infty} \right) 4\pi^2\nu_0^2\mathbf{w}_l \qquad (30)$$

Since $\mathbf{w}_t$ and $\mathbf{w}_l$ are periodic, with the transverse and longitudinal frequencies ν_t and ν_l, it follows that

$$\nu_t = \nu_0 \qquad (31)$$

and

$$\nu_l = \left(\frac{\epsilon_0}{\epsilon_\infty} \right)^{1/2} \nu_0 = \left(\frac{\epsilon_0}{\epsilon_\infty} \right)^{1/2} \nu_t \qquad (32)$$

Thus the dispersion frequency is identical with the transverse optical mode frequency, and the longitudinal frequency ν_l is given by the Lyddane–Sachs–Teller [26] formula (32).

In a nonionic crystal such as diamond or Ge, in the absence of polar interactions, the atomic motions are determined only by the local elastic restoring forces. Therefore, the second term of Eq. (12) vanishes. But since macroscopically E $\neq$ 0, it follows that

$$b_{12} = 0$$

and consequently

$$\nu_l = \nu_t = \nu_0 \qquad (32')$$

Since the macroscopic equations of motions are only valid for the long-wave limit, the above considerations apply only to the zone-center ($\mathbf{k}$ = 0) or the factor group modes.

Reststrahlen Spectrum

As a consequence of the infrared dispersion by ionic crystals, electromagnetic radiation with frequencies in the vicinity of the dispersion frequency undergoes selective reflection. For normally incident radiation in the case of an ideally ionic crystal, the reflectivity may be 100%, for which Eq. (20) holds good. This selective reflection of radiation in the neighborhood of the optical lattice mode frequencies in ionic crystals is known as the *reststrahlen phenomenon*. It can be understood by means of Eq. (20) along with the Fresnel formula

$$R = \left| \frac{n-1}{n+1} \right|^2 \tag{33}$$

where R is the fraction of light intensity reflected by an optically isotropic medium when light is incident perpendicular to its surface; $n = \sqrt{\epsilon(\nu)}$ is the refractive index and may assume complex values. It will be noticed from the dispersion relation (20) that at $\nu = 0$, $\epsilon(\nu) = \epsilon_0$, which is its static value. As ν increases, n increases steadily above the value $n = \sqrt{\epsilon_0}$. When ν reaches the dispersion frequency ν_0, ϵ, and hence n become infinite, and the crystal becomes perfectly reflecting with $R = 1$. When ν is further increased by an infinitesimal amount, $\nu_0^2 - \nu^2$ takes a negative but infinitesimally small value, making $\epsilon = -\infty$. With increasing ν, ϵ remains negative until it becomes zero again for a frequency satisfying the relation

$$0 = \epsilon_\infty + \frac{\epsilon_0 - \epsilon_\infty}{\nu_0^2 - \nu^2} \nu_0^2$$

the solution of which is $\nu = (\epsilon_0/\epsilon_\infty)^{1/2}\nu_0$. The refractive index $n = \sqrt{\epsilon}$, is therefore imaginary between the values ν_0 and $\nu = (\epsilon_0/\epsilon_\infty)^{1/2}\nu_0$, and $R = 1$ over this range. However these two values of ν are precisely the transverse and the longitudinal optical modes as may be seen from Eqs. (31) and (32). The reststrahlen band for an ideally ionic crystal is shown in Fig. 10 (top curve), for which a band of perfect reflection exists between the frequencies ν_t and ν_l.

In all real diatomic cubic ionic crystals, the observed reflectivity shows characteristic reststrahlen bands with high reflectivity between the frequencies ν_l and ν_t. However, in shape or intensity the observed reflectivity does not agree quantitatively with that of an ideal ionic crystal. This is evident in Fig. 11, where the reflection spectrum of AlSb is shown. This discrepancy is due to the fact that Eq. (20), though capable of representing dispersion of a real crystal at frequencies away from ν_0, is inadequate in the immediate vicinity

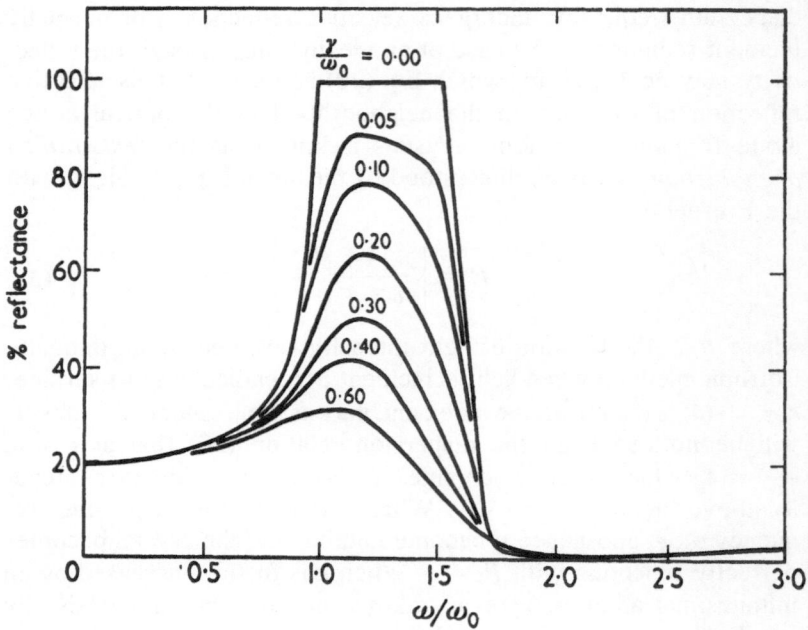

Fig. 10. Reflection spectra of a damped oscillator. (Reprinted by
permission from S. S. Mitra, Ref. 53.)

of the dispersion frequency. In real crystals, the strong reflection in
the reststrahlen region is also found to be associated with strong
absorption, whereas relation (20) predicts no such selective absorption.

A relation of the form of Eq. (20) was obtained from the phenomenological equations of motion (12) and (13). The latter equations neglected all but linear terms, with the consequence of mutu-
ally independent lattice waves. However, in real crystals they are
coupled by anharmonic and higher-order electric moment terms,
which play an important role in the dissipation of energy. Huang
[24] has shown that the energy density of the lattice waves is predominantly mechanical, manifested in the oscillations of the particles, and only a small portion is associated with the electromagnetic
field. Thus in the steady state, a small amount of energy dissipated
by the lattice waves due to a small amount of coupling between the
modes results in a drastic reduction of the electromagnetic energy
flux, with the consequent absorption of radiation by the medium.

A more realistic dispersion formula may be obtained by the

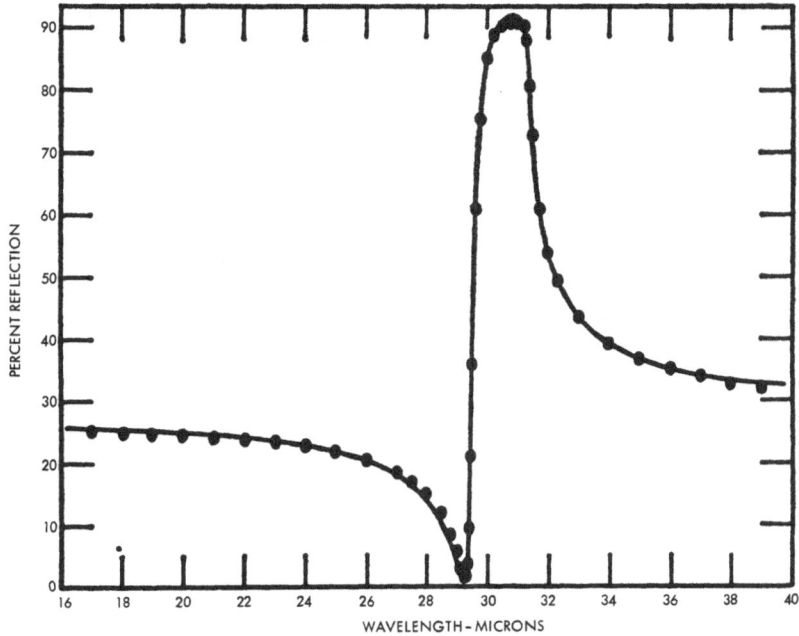

Fig. 11. Reststrahlen spectrum of AlSb. (Reprinted by permission from W. J. Turner and W. E. Reese, Ref. 29.)

phenomenological introduction of a damping term that represents a force which is always opposed to the motion and proportional to the velocity. This procedure provides a method for including the effect of energy dissipation in the neighborhood of ν_0. Equation (12) now takes the form

$$\ddot{\mathbf{w}} = b_{11}\mathbf{w} - \gamma\,\dot{\mathbf{w}} + b_{12}\mathbf{E} \qquad (34)$$

Introducing the periodic solutions $(\mathbf{w}.\mathbf{E}) = (\mathbf{w}_0, \mathbf{E}_0)e^{2\pi i\nu t}$, one obtains

$$-4\pi^2\nu^2\mathbf{w} = (b_{11} + 2\pi i\nu\gamma)\,\mathbf{w} + b_{12}\mathbf{E} \qquad (35)$$

Comparing Eq. (35) with Eq. (16), one notes that the addition of the damping term is equivalent to replacing the coefficient b_{11} by $b_{11} + 2\pi i\nu\gamma$, with the corresponding change in the dispersion formula (19). Therefore, Eq. (20) may be written as

$$\epsilon(\nu) = \epsilon_\infty + \frac{\epsilon_0 - \epsilon_\infty}{1 - (\nu/\nu_0)^2 - i(\gamma/2\pi\nu_0)(\nu/\nu_0)} \qquad (36)$$

which now includes absorption.

Now, a plane electromagnetic wave of phase velocity $c/\sqrt{\epsilon}$ $= 2\pi\nu/\mathbf{k}$ and frequency ν may be represented by

$$E = E_0 \exp 2\pi i\nu(\mathbf{e} \cdot \mathbf{r}(\sqrt{\epsilon}/c) - t) \qquad (37)$$

where $\mathbf{e}$ is a unit vector in the direction of the Poynting vector. In an absorbing medium the refractive index, and hence the dielectric constant $\epsilon(\nu)$, represent complex quantities given by

$$\sqrt{\epsilon(\nu)} = n + i\kappa \qquad (38)$$

Eq. (37) thus takes the form

$$E = E_0 \exp (-2\pi\nu\kappa \, \mathbf{e} \cdot \mathbf{r}/c) \exp \{2\pi i\nu \, (\mathbf{e} \cdot \mathbf{r} \, (n/c) - t)\} \quad (39)$$

and simultaneously describes the effects of refraction and attenuation. The real quantities n and κ are the refractive index and the extinction coefficient, respectively. The first term in Eq. (39) represents attenuation. In terms of the absorption coefficient α, the attenuation is given by

$$|E^2| = |E_0^2| \exp (-\alpha \mathbf{e} \cdot \mathbf{r}) \qquad (40)$$

Therefore, the relationship between α and κ is

$$\alpha = \frac{4\pi\kappa}{\lambda}$$

By expanding Eq. (36) in terms of its real and imaginary components, it follows that

$$n^2 - \kappa^2 = \epsilon_\infty + \frac{(\epsilon_0 - \epsilon_\infty) \, [1 - (\nu/\nu_0)^2]}{[1 - (\nu/\nu_0)^2]^2 + (\gamma/2\pi\nu_0)^2 \, (\nu/\nu_0)^2} \qquad (41)$$

and

$$2n\kappa = \frac{(\epsilon_0 - \epsilon_\infty) \, (\gamma/2\pi\nu_0) \, (\nu/\nu_0)}{[1 - (\nu/\nu_0)^2]^2 + (\gamma/2\pi\nu_0)^2 \, (\nu/\nu_0)^2} \qquad (42)$$

As a consequence of Eq. (38) the reflectance R of an absorbing medium is given by

$$R = \frac{(n - 1)^2 + \kappa^2}{(n + 1)^2 + \kappa^2} \qquad (43)$$

The reflection spectrum of a damped oscillator for several values of the damping factor γ/ω_0 (where $\omega_0 = 2\pi\nu_0$) is shown in Fig. 10.

Determination of the Optical Constants in the Reststrahlen Region

The refractive index n and the extinction coefficient κ are known as the optical constants of an absorbing medium. These can be determined as functions of frequency from the reflection spectrum by using Eqs. (41)–(43). If the values of static or low-frequency (~ 1 kc/s)ϵ_0 and n ($= \sqrt{\epsilon_\infty}$) in the visible or ultraviolet region free from any electronic transition are known, R may be calculated as a function of ν/ν_0 for several values of γ/ω_0. The inverse of the peak reflectivity is approximately linear with the damping constant [27]. Figure 12 shows a plot of $1/R_{max}$ versus γ/ω_0 for the damped oscillator of Fig. 10. From the observed value of the peak reflectivity, a value may be obtained for γ/ω_0. Values of n and κ may then be calculated from Eqs. (41) and (42). For a more accurate evaluation, Eqs. (41)–(43) are fitted to the observed

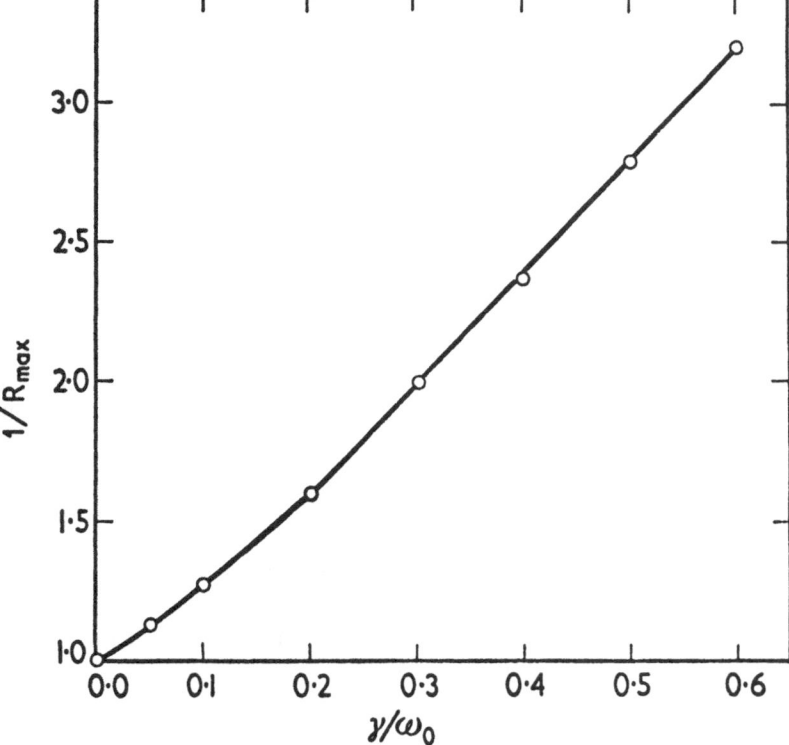

Fig. 12. Inverse of peak reflectivity *vs.* damping constant. (Reprinted by permission from S. S. Mitra, Ref. 53.)

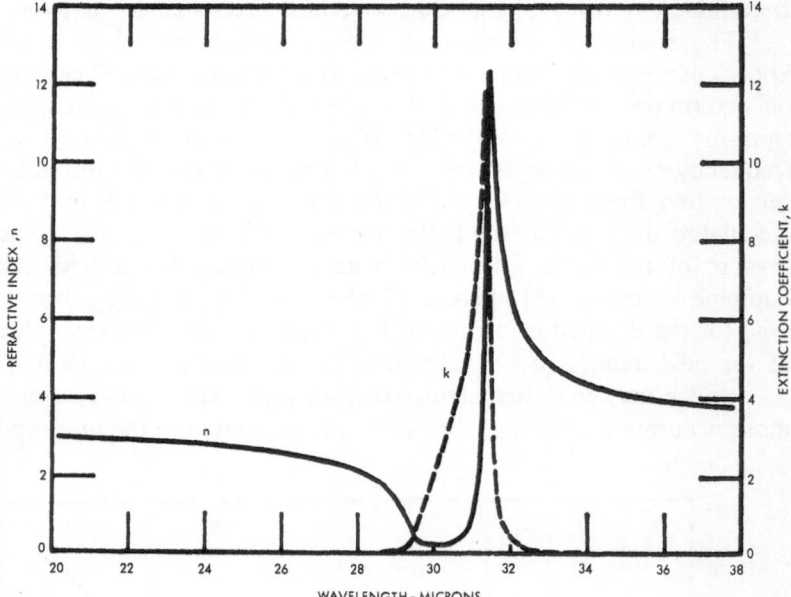

Fig. 13. Optical constants of AlSb. (Reprinted by permission from
W. J. Turner and W. E. Reese, Ref. 29.)

reststrahlen band by machine programming [28]. The best values
of ν_0, ϵ_0, ϵ_∞, and $\gamma/2\pi\nu_0$ thus obtained are used to calculate n and
κ. Figure 13 gives the calculated optical constants [29] for AlSb
from the reststrahlen spectrum shown in Fig. 11. In this figure the
best fit is given for the calculated values of R using the empirical
parameters. It may be mentioned here that although the maximum
of κ occurs very near $\nu = \nu_0$, it is $n\kappa\nu$ that undergoes the maximum
precisely at $\nu/\nu_0 = 1$.

The dispersion frequency ν_0 may also be determined directly
by transmission measurements. However, for a polar crystal the
strong absorption in the immediate vicinity of ν_0 necessitates the
use of very thin samples of the crystal. For extremely thin ($< 1\ \mu$)
samples, it may be shown that the minimum transmission occurs
exactly at the dispersion frequency.

Very often an elaborate determination of the optical constants
from the analysis of the reststrahlen band may not be necessary for
the evaluation of ν_0. It may be obtained from ν_m, the frequency at

which the maximum reflection occurs, provided ϵ_0 and ϵ_∞ are known. For this purpose Havelock [30] has shown that if γ/ω_0 is small, the ratio ν_m/ν_0 is approximately independent of the damping constant, and is given by

$$\frac{\nu_m}{\nu_0} = \left(1 + \frac{\epsilon_0 - \epsilon_\infty}{6\epsilon_\infty - 4} \right)^{1/2} \tag{44}$$

Since for the most ionic crystals the damping constant γ/ω_0 is relatively small, the above formula may be used for the determination of the dispersion frequency. This is also equal to the zone-center transverse optical mode frequency.

Under higher resolution the reststrahlen band is occasionally accompanied by some structure, usually a shoulder, on the high-frequency side. This side band corresponds to a second resonance frequency arising from the combination of an acoustic phonon with an optical phonon in accordance with the conservation principles discussed in the section on multiphonon absorption in single crystals. The analysis of such a reststrahlen band may be accomplished [28] by assuming two sets of parameters, one each for the two resonance frequencies, in Eqs. (41) and (42).

Determination of the Longitudinal Optical Mode Frequency at the Zone Center

The inactive zone-center longitudinal optical mode frequency ν_l of ionic crystals may be evaluated by several methods. Usually ν_l is computed from ν_t by using the Lyddane–Sachs–Teller relationship [Eq. (32)] provided ϵ_0 and ϵ_∞, the values of the dielectric constant at very low and high (visible) frequencies, are available. However, there may be considerable uncertainties regarding the values of ϵ_0 and ϵ_∞ to be used in many substances. This applies especially to crystals that are not good insulators, and to those with more than one reststrahlen band or those with low-energy electronic transitions. A second method of obtaining ν_l is to fit the observed reststrahlen spectrum using a dispersion relation in which ν_l is a variable parameter instead of ν_t. This method is not very accurate either, because dielectric dispersion relations need several undetermined parameters for a good fitting, and the data may be fitted quite accurately with some range of values of ν_l. Finally, ν_l may be determined by Drude's method, which consists of finding the frequency at which the real part of the dielectric constant (n^2) goes through zero. This may be accomplished by measuring reflectivity as a function of angle of incidence and direction of polarization. For strongly ionic crystals with a low value of the damping constant, for which Eq. (20) may be used as an approximate dispersion relation, ν_l corresponds to the frequency at which the reflectivity attains a minimum on the high-frequency side of the reststrahlen band.

Inelastic neutron scattering studies may yield dispersion curves for the longitudinal optical mode frequency as a function of the wave vector $\mathbf{k}$. A fairly accurate value of ν_l can be obtained from the extrapolation of these curves to $\mathbf{k} = 0$. Brockhouse and his associates have obtained ν_l for several alkali halide crystals by this method [31]

Finally, Berreman [32] has recently shown that thin films of cubic ionic crystals have sharp, strong infrared absorption and reflection at frequencies characteristic of the longitudinal optical mode of long wavelength when the radiation is p-polarized and incident obliquely to the surface.

Analysis of Reflection Spectra Due to Internal Modes

Certain infrared-active internal fundamentals of some crystals may also be very intense. This makes it necessary to use extremely thin ($\sim 1\mu$) samples in order to obtain satisfactory transmission data. With available techniques, however, it is well-nigh impossible to prepare single crystals of this thickness in various orientations for many crystals. Auxiliary information on the absorption in such situations may be available from the reflectivity, which is necessarily large because of anomalous dispersion associated with the large extinction coefficient. It is possible to relate the observed reflection spectrum to the optical constants near the dispersion frequency by means of Kramers–Kronig analysis.

The maximum reflection associated with an infrared-active internal mode seldom exceeds 50%, and is usually within 20%. The method described in an earlier section for the analysis of the lattice reststrahlen band is therefore not suitable for the determination of the optical constants in the vicinity of an internal mode frequency of a molecular crystal. In such cases n and κ may be obtained from measurements of reflection spectra at two widely separated angles of incidence. Simon [33] has reviewed the experimental procedure and the method of calculation. This method however works only for isotropic solids. Since the extinction coefficient is a function of the angle of incidence for anisotropic crystals, the method is rendered unsuitable. Robinson and Price [34] have modified Šimon's method to permit the evaluation of the optical constants in the neighborhood of a strong absorption band from the measurement at normal incidence. They find that n and κ are given by

$$ n = \frac{1 - r^2}{1 + r^2 - 2r \cos \theta} \qquad \kappa = \frac{- 2r \sin \theta}{1 + r^2 - 2r \cos \theta} \qquad (45) $$

The quantity $R = r^2$ is the experimentally observed reflectivity. The phase difference θ between the incident and the reflected waves is obtained from

$$ \theta_c = \frac{1}{\pi} \int_0^\infty \frac{d\ln r}{d\nu} \ln \left| \frac{\nu + \nu_c}{\nu - \nu_c} \right| d\nu \qquad (46) $$

where θ_c is the value of the phase difference at a frequency ν_c. The integral is evaluated numerically. The limits of integration are, in practice, the two extremities of the band where the reflection approaches constant values.

The infrared transmission in the hydroxyl stretching region of single crystal brucite, $Mg(OH)_2$, is very low for certain orientations of the crystal and polarizations of incident radiation. This is especially true for the absorption near 3700 cm^{-1} for radiation polarized parallel to the c axis. The absorption is too high, rendering the accurate determination of its band position very difficult. The associated high reflectivity in this region is shown in Fig. 14 for near normal incidence on a plane cut perpendicular to the cleavage plane. The only reflection band has a maximum at 3690 cm^{-1}. It is completely dichroic, with the infrared transition moment vector oriented along the c axis of the crystal. The optical constants obtained from an analysis of this band are given as functions of frequency in Fig. 15. The peak of the extinction coefficient curve occurs at 3705 cm^{-1} and its value is 0.57. Thus for a sample of 1-μ thickness, the maximum absorption at this frequency will be 94%, which explains the difficulty encountered in the measurement of single-crystal transmission spectra. The transmission spectrum of a powdered sample gives a value of 3700 cm^{-1} for the position of this absorption band in excellent agreement with the value obtained from the analysis of the reflection spectrum.

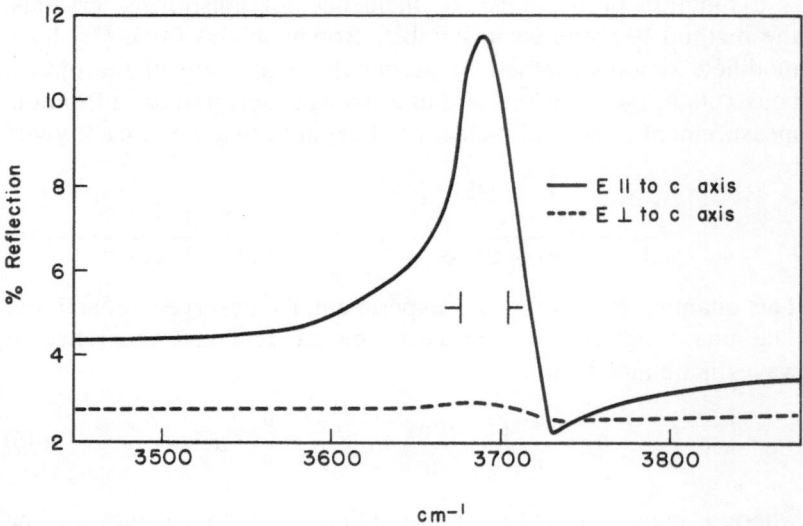

Fig. 14. Reflection spectrum of brucite from 3450 to 3850 cm⁻¹. (Reprinted by
permission from S. S. Mitra, Ref. 1.)

The Robinson – Price method of analysis is rigorously applicable only to crystals of orthorhombic or higher symmetry. Under
certain conditions however the method may also be applicable to
crystals of lower symmetry. This method may give erroneous results
for ferroelectric solids. The use and applicability of the Robinson –
Price method has recently been reviewed and critically examined
by Schatz, Maeda, and Kozima [35] and by Bowlden and Wilmshurst [36].

Attenuated Total Reflection
It has been noted in the foregoing that reflection spectral measurements are in general useful for two reasons. First, the analysis
of the reflection spectrum affords a means of obtaining the optical
constants as function of frequency, and second, it permits spectral
measurements on an otherwise intractable sample, such as a highly
absorbing crystal, for which it is impractical to prepare a specimen
of required thinness. The reflection technique, however, is limited
to the regions for which the extinction coefficient of a substance has
a high value. For example, crystals with extinction maxima in the

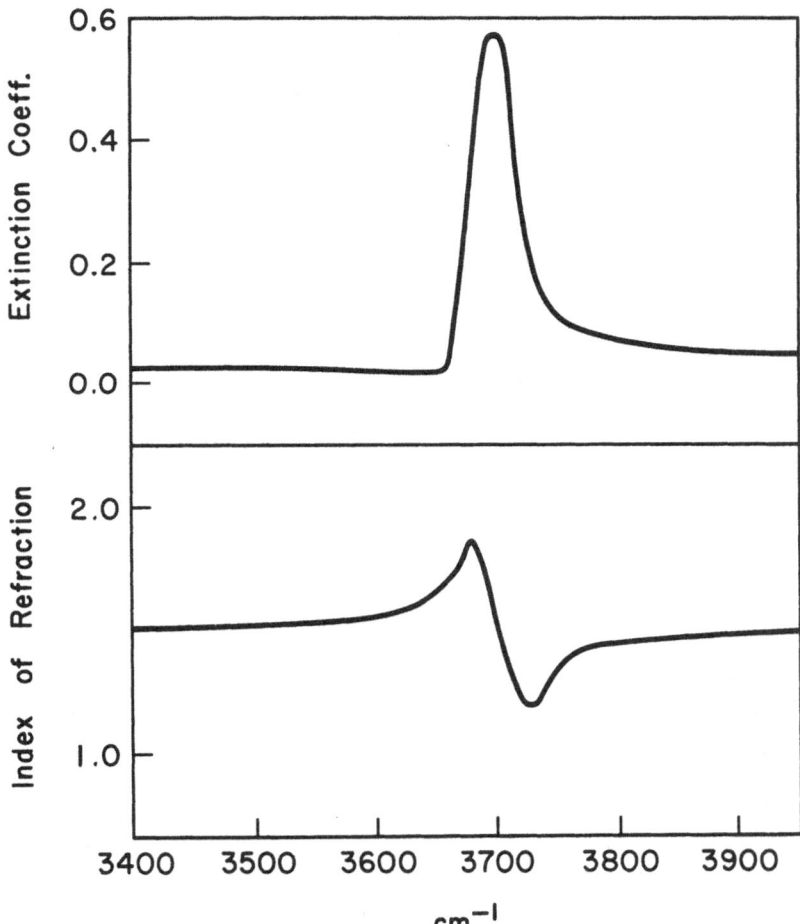

Fig. 15. Optical constants of brucite in the region 3400–3900 cm⁻¹. (Reprinted by permission from S. S. Mitra, Ref. 1.)

range of 0.0–0.2 will show reflection spectra of low overall intensity. They will be devoid of contrast, to render the determination of the optical constants with any accuracy impossible. Since a large number of absorption bands associated with the vibrational modes of a great number of organic and nonpolar solids are weak, the ordinary reflection technique gives very little useful information for such materials.

The above difficulties can be overcome by a novel reflection technique developed by Fahrenfort [37], known as attenuated total reflection (ATR). This technique utilizes the interface between a dielectric of high refractive index and the specimen as the reflecting surface, instead of that between air and the sample used in the conventional technique. In the new method the refractive index n at the interface will thus be less than 1. If the angle of incidence θ and the dielectric are selected in such a way that $n \leq \sin \theta$, total reflection ensues for wavelengths where the sample is nonabsorbing. If, however, for the second medium $\kappa \neq 0$, part of the incident radiation will be absorbed by the surface layers, reducing the reflected energy. It may be shown that significant reflection attenuation takes place even for very small values of the extinction coefficient. The attenuated reflection spectrum is in many ways similar to a transmission spectrum, because with increasing extinction coefficient the attenuation increases, progressively decreasing the light intensity leaving the sample. Optical constants may also be determined from such spectra with a slightly different Kramers – Kronig analysis than described earlier.

KRS-5, AgCl, or Ge single crystals seem to be suitable dielectrics for ATR measurements. These crystals are transparent and have refractive indices between 2 and 4 through considerable part of the infrared region, and thus may be used to produce interfaces of $n < 1$ with any absorbing specimen with refractive index less than 2. Figure 16 shows the comparison between the conventional reflection spectrum and the ATR spectrum of a solid epoxy resin.

Emission Spectra in the Infrared

Electromagnetic radiation incident on a solid is partly reflected, partly absorbed, and partly transmitted. If all the incident radiation is absorbed, the solid is termed a *blackbody*. If the solid is at a higher temperature than the environment it also possesses an emission spectrum. The thermal radiation emitted by a blackbody follows Planck's law. Emission from a solid which is not a perfect absorber is less than that from an ideal blackbody at the same temperature. The two quantities are related by Kirchhoff's law.

For partially absorbing solids the emissivity is related to the reflectance and the transmittance. By considering multiple reflections in a plane-parallel sample, McMahon [38] has derived expressions for the apparent reflectivity, the apparent transmissivity, and

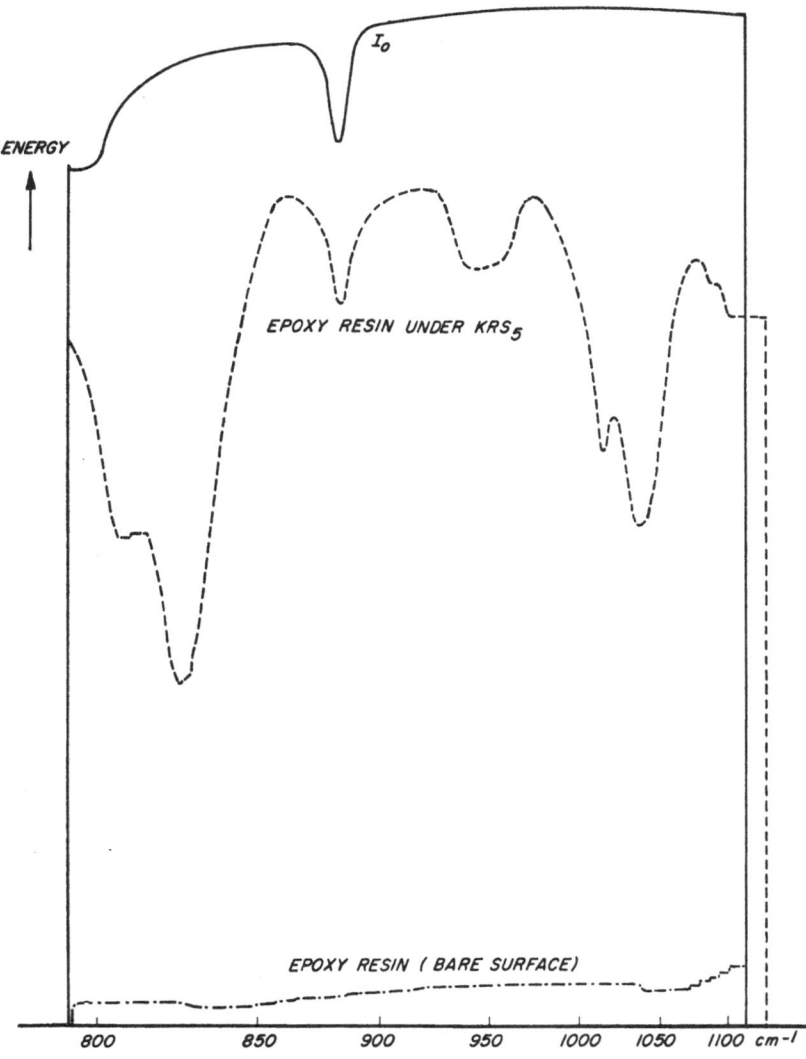

Fig. 16. Comparison between conventional reflection spectrum and ATR spectrum of a solid epoxy resin. (Reprinted by permission from J. Fahrenfort, Ref. 37.)

the emissivity, in terms of the true reflectivity and the true transmittance. The true reflectivity is the fraction of incident radiation that is reflected from the first surface, and is given by the Fresnel formula (33). The true transmittance is the fraction of light entering the solid that reaches the second surface, i.e., without any internal reflection, and is given by

$$T = e^{-\alpha t} \tag{47}$$

where $\alpha = 4\pi\kappa/\lambda$ is the absorption coefficient, and t is the thickness of the sample. McMahon's relationships for the apparent, or observed, quantities are

$$R^* = R \left[1 + \frac{T^2 (1 - R)^2}{1 - R^2 T^2} \right] \tag{48}$$

$$T^* = T \frac{(1 - R)^2}{1 - R^2 T^2} \tag{49}$$

and

$$E = \frac{(1 - R) (1 - T)}{1 - RT} \tag{50}$$

By adding Eqs. (48), (49), and (50) one obtains

$$R^* + T^* + E = 1 \tag{51}$$

Haas [39] has measured R^*, T^*, and E as functions of frequency for $CaCO_3$ and $NaNO_3$, and has found that the relation (51) is strictly obeyed. At first glance it appears from Eq. (51) that no new information is available from emission measurements which can not be obtained from reflection or transmission measurements. However, closer scrutiny reveals that for certain spectral regions emissivity measurements can be very effective in acquiring optical data which are difficult to obtain by transmission or reflection measurements. Substitution of Eq. (47) in Eq. (50) gives

$$E = \frac{(1 - R) (1 - e^{-\alpha t})}{1 - Re^{-\alpha t}} \tag{52}$$

In the case of an almost transparent sample with very low absorption coefficient ($\alpha t << 1$), Eq. (52) reduces to

$$E \simeq \alpha t \tag{53}$$

Direct measurement of the emittance thus offers a distinct advantage over reflectance and transmittance measurements. For low values of αt, the emissivity measurement is a sensitive method of determining α. Using a nominal sample size, Stierwalt and Potter [40] have measured extremely low absorption coefficients of Si, Ge, and CdS. The determination of the absorption coefficient of these materials by the transmission method would have required very thick samples in the regions where the absorption coefficient is as low as 0.2 cm⁻¹. The emission technique therefore seems eminently suitable for the study of absorbance in the overtone and combination regions. Figure 17 gives the spectral emittance of a 1.7-mm-thick single crystal of silicon at several temperatures.

For almost opaque solids, on the other hand, the absorption coefficient and hence αt for nominal sample thickness are large, so that $T \simeq O$ and $R^* \simeq R$. The emissivity is therefore given by

$$E = 1 - R \qquad (54)$$

and the direct measurement of E permits an accurate determination of small changes of the reflectance.

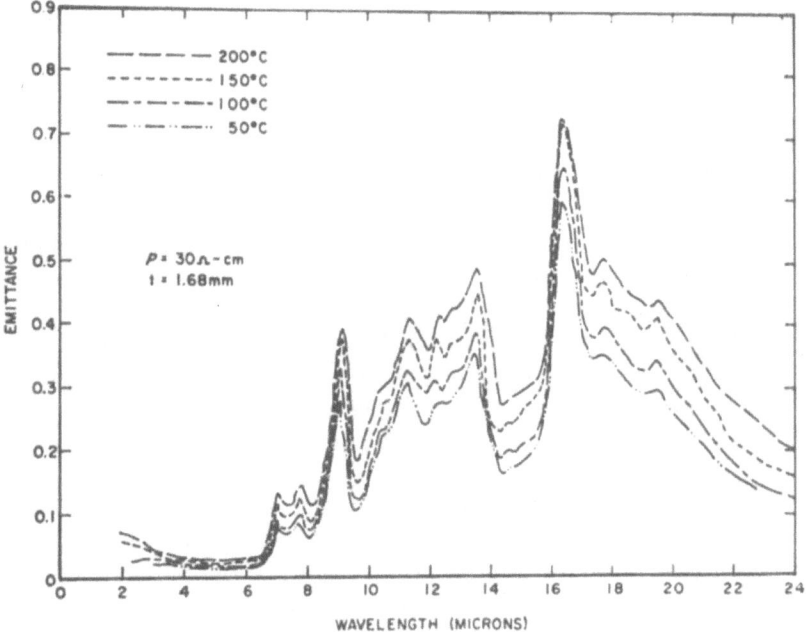

Fig. 17. Spectral emittance of single-crystal Si as a function of temperature. (Reprinted by permission from D. L. Stierwalt and R. F. Potter, Ref. 40.)

COMBINATIONS AND OVERTONES

In the harmonic approximation, only the fundamental modes of vibration in a solid are possible. Combinations and overtones do not arise. Moreover in the infrared absorption or Raman scattering, the fundamentals should appear as sharp lines. Their optical activity depends on whether or not the equivalent atoms move identically in phase. If θ_{ij} is the phase difference for a mode between the ith and jth unit cell, the condition that the mode be optically active is that $\theta_{ij} = 0$, which corresponds to the modes that belong to the center of the Brillouin zone with $\mathbf{k} = 0$. The appearance of combinations and overtones in the spectra, the line broadening in the condensed phases, and the phenomenon of thermal expansion of solids, however, indicate the presence of cubic or higher-order terms in the vibrational potential energy function of a solid. Inclusion of anharmonicity in the general theory of the vibrational spectra of crystals not only allows combinations and overtones but puts very little restriction on the selection rules applicable to such transitions.

Multiphonon Absorption in Simple Crystals

From elementary theory, the infrared absorption spectrum of a diatomic ionic crystal like NaCl or ZnS is expected to consist of a single line associated with the transverse optical lattice mode of essentially zero propagation vector. For diatomic homopolar crystals like diamond or silicon this factor group fundamental should be a single line active only in the Raman scattering. However, in reality, the resonance absorption is not infinitely sharp but shows the natural line width due to radiation damping. The excited state may also transform through some coupling mechanism into a state in which two or more different phonons replace the optical one. This process also limits the lifetime of the optical phonon causing broadening and/or side bands on the short-wavelength side.

For polar crystals the mechanism of interaction of the electromagnetic field with phonons causing the side bands has been explained [41] as due to the anharmonic part of the potential energy associated with the lattice vibrations. The interaction takes place through the dipole moment associated with the lattice mode. In homopolar crystals the factor group fundamental is inactive because the dipole moment is zero, and thus the above mechanism does not apply. For such systems, Lax and Burstein [42] have shown that two or more phonons can interact directly with the radiation through terms in the electric moment of second or higher order in the atomic displacements. Such combinations, of either origin, will be optically active if the energy and wave vectors are conserved according to

$$hv = \sum_i (\pm hv_i) \tag{55}$$

and

$$\mathbf{q} + n\mathbf{K} = \sum_i (\pm \mathbf{k}_i) \tag{56}$$

where hv and $\mathbf{q}$ are the energy and the wave vector of the absorbed photon, hv_i and $\mathbf{k}_i$ are the energy and the wave vector of the ith photon, $\mathbf{K}$ is a reciprocal lattice vector, and n is an integer. The positive and negative signs indicate emission and absorption, respectively. Since the wave vector of the photon in the infrared region is small compared to the phonon wave vector, the second condition becomes

$$n\mathbf{K} = \sum_i (\pm \mathbf{k}_i) \tag{57}$$

For one- and two-phonon processes $n = 0$, and for a three-phonon process $n = 0, \pm 1$. In the event of interaction of the photon with a single phonon, the latter must be a zero wave vector phonon, i.e., the factor group fundamental that belongs to the zone center. A two-phonon sum band requires $\mathbf{k}_1 = -\mathbf{k}_2$ as a consequence of Eq. (57), while for a difference band the requirement is $\mathbf{k}_1 = \mathbf{k}_2$.

Strictly speaking, the combinations should give rise to continuous absorption. Absorption maxima, however, occur because of singularities in the phonon frequency distribution. Only those phonons in each branch which arise from regions where there is a high density of phonon states per unit wave vector interval participate in optical processes. These regions, or points of high phonon concentration, in the phonon dispersion, are known as critical points (c.p.). The singularities corresponding to the critical points occur where the dispersion curves for the individual branches are flat. So, in general, if there are p c.p.'s in the Brillouin zone for a crystal, then, barring residual degeneracies, as many as $3np$ indi-

vidual phonons may take part in the combination spectrum, where
$3n$ are the number of branches corresponding to n particles per unit
cell. Each of these phonons may be assigned to one of the irreducible representations of the crystal space group (and not the factor
group), or, in other words, it must belong to a symmetry species of
the crystal. It is possible to derive the optical activity and the
selection rules governing the combinations among such a system of
crystal modes, starting from the space group of the crystal. The
selection rules for the multiphonon combinations among the c.p.
phonons have been worked out from space group theory by Birman
[43] for the diamond and the zinc blende type of structures. The application of these to infrared spectra, however, is not obvious because of the lack of complete dispersion relations. The observed
infrared spectra of diamond, AlSb, ZnS, etc. are usually analyzed
phenomenologically by a four-phonon scheme. These characteristic
phonon frequencies are assumed to belong to the Brillouin zone
boundary, where the dispersion curves for the individual branches
are flat, resulting in singularities in the phonon density. A given
phonon branch is thus characterized by a single phonon frequency.
The broadening of the combination bands is attributed to the strong
dependence of the energy on the wave vector direction near the
Brillouin zone boundary. Thus for the diatomic cubic crystals
(NaCl, CsCl, diamond, and the zinc blende type), one assumes four
characteristic phonons: LO, TO, LA, and TA, corresponding to the
zone edge values of the longitudinal and transverse optical and
acoustic branches. Occasionally, a breakdown of the twofold
degeneracy of the transverse branches along the different directions
in $\mathbf{k}$ space is assumed. This results in additional characteristic
phonon frequencies which are needed for the empirical analysis of
the observed spectra.

The temperature dependence of the absorption spectrum at any
frequency of incident radiation depends on the number of phonons
available to take part in a proposed multiphonon process at a given
temperature. The temperature dependence of the number N_i of
phonons of frequency ν_i is given by

$$N_i = [\exp(h\nu_i/kT) - 1]^{-1} \tag{58}$$

The probability [44] of absorption of one of these phonons is proportional to N_i, whereas the probability of the emission of such a
phonon is proportional to $1 + N_i$. Let us suppose an absorption

band is assigned to a two-phonon process such that the absorption of a photon is accompanied by the emission of two phonons ($i = 1,2$) in accordance with the conservation principles (55) and (56). The probability of the process is proportional to $(1 + N_1)(1 + N_2)$. The net absorption is obtained by correcting for spontaneous emission of the photon by the reverse process in which two phonons are absorbed. Then the temperature dependence of the net absorption is proportional to $(1 + N_1)(1 + N_2) - N_1 N_2 = 1 + N_1 + N_2$. Similarly, for an assigned three-phonon process such as $LO(\nu_1)$ + $TO(\nu_2) - LA(\nu_3)$, the absorption coefficient will be proportional to $(1 + N_1)(1 + N_2) N_3 - N_1 N_2 (1 + N_3)$. The two conservation laws plus the temperature dependence of the absorption intensity subject the phenomenological analysis of the spectral data to rather stringent self-consistency requirements.

Combinations and Overtones in Crystals Containing Molecular Groups [45]

It was observed that the fundamental modes of vibration that may be infrared or Raman active belong to the center of the Brillouin zone. They may be subdivided into internal, translatory external, and rotatory external modes in the case of crystals containing molecular groups. Their symmetry classifications and selection rules are derived by factor group analysis, and these constitute the $k = 0$ modes of the optical branches. The acoustic modes have zero frequencies in the factor group approximation and hence are excluded. However, the latter will have nonvanishing values at the zone boundary, and thus may take part in combination tones.

In general, there may be the following types of combinations in a molecular crystal:

Internal + Internal	(59)
Internal + External	(60)
Internal + Acoustic	(61)
External + External ⎫	
External + Acoustic ⎬ Multiphonon	(62)
Acoustic + Acoustic ⎭	

The only selection rules that may be operative are the two conservation laws (55) and (56). Whereas the interactions (59), (60), and (61) involve molecular modes, the last three interactions denoted by

(62) are formally equivalent to the multiphonon combinations discussed in the earlier section.

For a crystal such as brucite, $Mg(OH)_2$, with five atoms per unit cell, there are 15 distinct branches. Ten of these branches correspond to optical lattice modes, three to the acoustic modes, and two to the internal modes [46]. In the simplest case of only one c.p. corresponding to $k/k_{max} = 1$, in addition to the multiphonon processes, one needs to consider combination of two internal mode frequencies with the 13 external mode frequencies. It may be emphasized that these combining frequencies are not the observed fundamentals (which are for $k = 0$), but their values at the zone boundary. Furthermore, any degeneracy (accidental or otherwise) will probably break down at the zone boundary making all these combining frequencies distinct. In a hexagonal crystal like $Mg(OH)_2$ (space group $P\bar{3}m$), it is highly unlikely to have only one c.p. One would expect a minimum of two c.p.'s (along the a and c axes of the crystal), and in all probability more. Thus, in reality, it may be necessary to consider the combination of four internal modes with 26 external modes, or perhaps six with 39, etc. It is needless to say that such a spectrum is expected to be very complex. The envelope shape of the combination bands will depend on the dispersion relations and may be worked out only if the latter are known. However, to our knowledge, the lattice dynamics has not yet been completely worked out for any molecular crystal.

For real molecular crystals with strong chemical bonds within a molecular group, with loose bindings such as the van der Waals type between the groups, there may be some simplification possible. This applies to combinations involving only the internal modes of the molecular groups. Usually in such crystals the interactions between the identical groups within a unit cell give rise to the so-called correlation field splitting discussed earlier, but interaction between groups in different unit cells may be neglected. Consequently, the k dependence of the internal mode frequencies may be limited, with the result that the zone-boundary frequencies may not be very different from the zone-center fundamentals. In such a situation, combination tones may be expected of the type observed in molecular spectra in the gaseous phase. In addition, if the molecular groups are heavy and have very loose binding between them, the phonon frequencies may be expected to be quite small and may be manifest only in the envelope shape of the internal fundamentals and their combinations. A situation like this may be expected in a crystal like naphthalene or anthracene.

In the electronic spectrum of solid N_2, Vegard [47] observed spectral spacings of 40 and 69 cm^{-1}, which were assigned to the excitation of librational modes. The infrared absorption spectrum of crystalline N_2 also shows this structure [48], and has been explained as arising from a libration–vibration combination. However, Ewing and Pimentel [49] from a consideration of the heat of sublimation (using an r^{-6}, r^{-9} potential function) have shown that the broad absorptions of solid N_2 may in part at least be caused also by combinations involving translations. Similar combinations have also been observed by them in solid carbon monoxide. In the infrared absorption spectra of solid N_2O and CO_2, using thick samples, Dows [50] has observed several combination bands. These he assigned as originating from the mixing of the libration modes with the bending mode of an XY_2 molecule. Hydrogen is one of the few molecules that undoubtedly shows free rotation in the solid state because of the extremely weak intermolecular forces. It is interesting to note, however, that the same intermolecular forces, along with quadrupolar interactions, are able to induce fundamental infrared absorption in the solid and the liquid states, in spite of the fact that such absorption is forbidden for the free molecule. The fine structure of the vibration spectrum of crystalline H_2 investigated by Gush *et al.* [51] can only be explained by the presence of a more or less unhindered rotation. Additional broad absorption features are assigned to a combination of the stretching mode with translational modes.

It has been a frequent practice to use the factor group selection rules also in the analysis of the combination spectra of the molecular modes. Unfortunately, theoretical justification exists only for application to the fundamentals, and not to such an extension. We shall use the gypsum crystal to illustrate this point. Consider the band at 2112 cm^{-1}, which seems numerically to be arising from the combination of the two sulfate internal modes $\nu_3 (B_g) = 1117$ cm^{-1} and $\nu_1 (A_u) = 1000$ cm^{-1}. Now, from the character table given in Table II, one finds the product

$$A_u \cdot B_g = B_u \qquad (63)$$

It is then argued that the combination band should manifest the symmetry properties of the B_u species, which are the infrared activity and transition moments along the $b(C_2)$ axis, i.e., in the (010) plane. However, these conclusions may not necessarily be correct because of the following reasons. First, the factor group selection rules may not be operative. A combination band judged inactive by the factor group selection rules, may as a matter of fact, be active. Second, the numerical agreement is no guarantee, because such an agreement also suggests a harmonic force field. However, the occurrence of a combination mode requires an anharmonic field. Finally, the combining frequencies may correspond to any **k** value, and may be different from the observed **k** = 0 fundamentals.

If the lattice modes in a crystal are substantially large, as in the case of light molecular groups with relatively strong binding, the analysis of the combination modes arising from interactions with the internal modes is not straightforward. For example, in the alkali and alkaline earth hydroxides, which are predominantly ionic with relatively light groups, the lattice modes range from 100 to 600 cm^{-1}. The combination of the internal with the lattice modes does indeed give an immensely complicated near-infrared spectrum [52], which can hardly be understood from a factor group standpoint.

To our knowledge brucite, $Mg(OH)_2$, is the only ionic molecular crystal for which the multiphonon combination bands and the importance of the zone-boundary acoustic modes have been pointed out [53]. For a complete understanding of the internal – external combination spectra of crystals containing polyatomic groups it may be necessary to know the dispersion of individual branches (from elastic-constant, specific-heat, or neutron-scattering data, etc.) and the location of the critical points in the Brillouin zone. In addition, high-resolution experimental data at several temperatures in the near- as well as in the far-infrared regions and Raman-scattering studies may be necessary. In principle the selection rules for combination modes can be worked out from group-theoretical considerations involving the entire space group. It appears to us, however, that the quantitative study of vibration spectra of solids is still in its infancy, and the present state of the art does not yet permit a thorough analysis for crystals of relatively complex structures containing even the simplest of the polyatomic groups. Even the shape of the combination bands of a simple crystal like Si is not predictable with much success [54]. This situation exists with the lattice dynamics and experimental dispersion curves being fairly well known. Thus, it is still a far way before the envelope shape of the combination bands of a crystal like $Mg(OH)_2$ (whose elastic constants are not even known) can be predicted on a realistic basis. It still remains to advance specific mechanisms which make the interactions of the internal and external modes in the hydroxide crystals possible. Hexter and Dows' model [55] and the modifications proposed by Mitra [56] of a librating vibrator may be regarded as attempts to construct such a mechanism. However, they incorporate only a portion of the possible interactions, namely, the coupling of the internal with the rotatory lattice modes. The interaction of the internal with the external modes should be fairly universal, and should be observed in varying degrees in most crystals containing polyatomic groups. High-resolution spectroscopy in transmission and in emission may be the key to revealing the interaction.

Buchanan [57] has reported complex near-infrared spectra for NaOH and KOH under certain conditions, whereas Snyder *et al.* [58] have failed to do so. The importance of impurities and point defects in accentuating forbidden transitions in Si and Ge is fairly well known [59]. One therefore suspects that the lattice – internal and phonon – phonon combinations in molecular crystals may also be affected by impurities and defects. This probably explains the difference between the spectra of Buchanan and those of Snyder for NaOH and KOH.

The vibrational combination spectrum (whether among lattice modes, between internal and lattice modes, or among internal modes) of solids is a highly complex situation. The concepts related to an isolated molecule (in the gas phase or dilute solution) or factor group considerations should be applied with caution to such spectra, and at best may be regarded as a very poor approximation. An almost infinite number of frequencies is possible in a crystal corresponding to a normal mode for which the vibrations in successive unit cells are out of phase by varying amounts. Each mode becomes active in combination with another mode of the same or of a different vibrational frequency having a corresponding compensating phase difference.

EXAMPLES

In this part, the optical vibration spectra of a few typical classes of crystals will be described in some detail, with a view to illustrating some of the principles discussed earlier. We will start with an example of a diatomic cubic crystal, aluminum antimonide. Next, gypsum, a semi-ionic molecular crystal, will be discussed, and finally the infrared spectrum of cyclopropane, a truly molecular crystal with van der Waals type binding, will be presented.

Except for certain diatomic crystals, we will not attempt to present a complete bibliography of all published experimental work on the infrared and Raman spectra of solids. It is felt that extensive bibliographies already exist [2, 60] for polyatomic crystals. In recent years, considerable interest has been evidenced in the study of the lattice infrared spectrum of semiconducting compounds of simple structure. In the following sections a review of the multiphonon spectra of such crystals will be presented.

Multiphonon Infrared Absorption in AlSb

AlSb belongs to the zinc blende type of structure with two atoms per Bravais unit cell. The lattice infrared spectrum of AlSb has been measured by Turner and Reese [29]. The reststrahlen spectrum and its analysis is given earlier. The maximum of the extinction coefficient occurs at $31.37 \pm 0.5 \mu$ (319 cm^{-1}), which is identified with the zone-center transverse optical mode, ν_t. The zone-center longitudinal optical mode ν_l is infrared inactive but may be identified with the short-wavelength minimum of the reflection spectrum at 29.4μ. A more precise value may be obtained from the Lyddane–Sachs–Teller relation [26], $\nu_l = (\epsilon_0/\epsilon_\infty)^{1/2} \nu_t$, already derived. The values of ϵ_∞, ϵ_0, and ν_t are obtained from the dispersion analysis of the reststrahlen band. The calculated value of ν_l is 340 cm^{-1}.

The absorption coefficient of AlSb as a function of wavelength is shown in Fig. 18. It was obtained from several transmission spectral measurements. The absorption peaks are due to the lattice absorptions involving one, two, three, and four phonons. The strongest absorption, in the region of 31μ, is due to the zone-center transverse optical mode phonon ν_t, the exact position of which was determined from an analysis of the reststrahlen band. The position of the zone-boundary characteristic phonons are estimated from the seven two-phonon band assignments as indicated in Fig. 18. The triplet 2LO, LO + TO, and 2TO is better resolved at 77°K. The assignment is confirmed by a comparison of the net power absorbed in a two-phonon summation at two temperatures.

A temperature factor f is defined as the ratio of the magnitudes

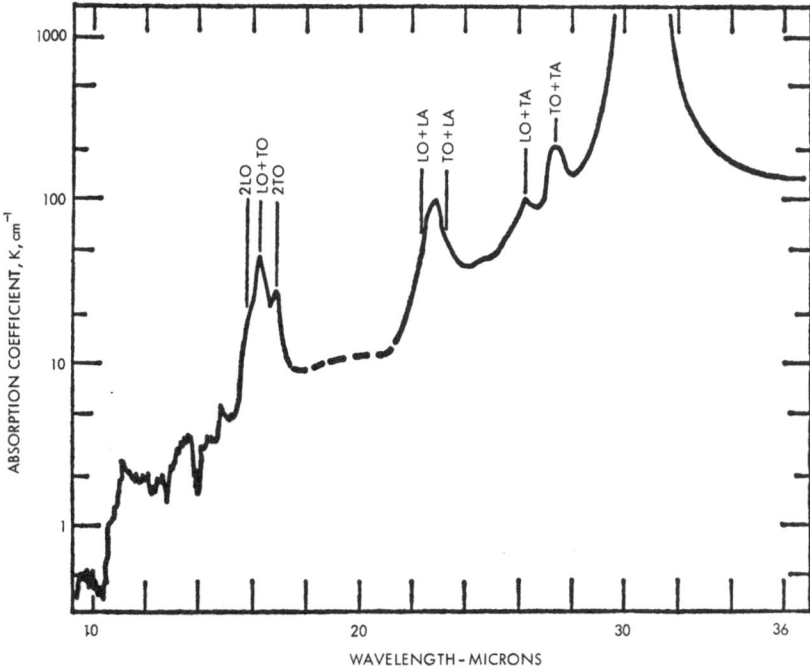

Fig. 18. Absorption spectrum of AlSb. (Reprinted by permission from W. J. Turner and W. E. Reese, Ref. 29.)

of the lattice band at 300°K and 77°K. It may be seen from our discussion of multiphonon absorption in simple crystals that the calculated temperature factor f_{cal} is given by

$$f_{cal} = \frac{(1 + N_1 + N_2)_{300°K}}{(1 + N_1 + N_2)_{77°K}} \qquad (64)$$

The observed temperature factor f_{obs} was obtained after correction for the background due to free carrier absorption, impurity absorption, and the tail of the fundamental absorption at ν_t. The assignment for the two-phonon bands along with their temperature dependence is given in Table V. Turner and Reese's assignment of the two optical phonons is quite arbitrary. As we shall note in what

TABLE V
Two-Phonon Combination Bands in AlSb as Calculated from the Four Characteristic Zone Boundary Phonon Frequencies: LO = 316 cm^{-1}, TO = 297 cm^{-1}, LA = 132 cm^{-1}, and TA = 65 cm^{-1a}

λ, μ	Observed ν,cm^{-1}	f_{obs}	Assignment	Calculated ν,cm^{-1}	f_{cal}
27.50	363	2.83	TO + TA	362	2.84
26.30	380	2.86	LO + TA	381	2.82
23.29	429	2.27	TO + LA	429	2.23
22.33	448	1.91	LO + LA	448	2.20
16.81	595	1.70	2 TO	594	1.62
16.31	613	1.57	LO + TO	613	1.59
15.80	633	1.56	2 LO	632	1.55

^{a}Reprinted by permission from W. J. Turner and W. E. Reese (Ref. 29).

follows, for a crystal of low ionicity such as AlSb it is preferable to designate the highest zone-boundary phonon as TO, and the next highest as LO. The assignment of the combination bands, however, is not otherwise affected. Thirty-eight additional bands were observed in the short-wavelength side of the 2LO band at 15.8 μ. These were assigned as three- and four-phonon combination bands, as shown in Table VI.

A value of the effective ionic charge q^* may be obtained from the relation

$$q^* = 2\pi\nu_t \left(\frac{\epsilon_0 - \epsilon_\infty}{4\pi} \right)^{1/2} \left(\frac{3}{\epsilon_\infty + 2} \right) \left(\frac{\mu\nu_a}{Ze} \right)^{1/2} \qquad (65)$$

due to Szigeti [61], where μ is the reduced mass per unit cell of volume ν_a, and Ze is the formal charge on an ion. For AlSb, $q^* = 0.48$, indicative of the fact that it is only partially ionic.

On basis of Coulomb attractive forces and nearest-neighbor repulsive forces, Brout [62] has given the sum rule

$$\sum_{i=1}^{6} \omega_i(\mathbf{k})^2 = \frac{18r_0}{\chi\mu} \qquad (66)$$

where $\hbar\omega_i(\mathbf{k})$ is the phonon energy of the ith vibrational branch at wave vector $\mathbf{k}$, r_0 is the nearest-neighbor distance, and χ is the com-

TABLE VI
Summary of the Three- and Four-Phonon Combination Bands in AlSb[a]

Line	No.	Assignment	Expected ν, cm^{-1}	Expected λ, μ	Observed ν, cm^{-1}	Observed λ, μ
	1	3LO + LA	1080	9.26	1081	9.26
	2	2LO + TO + LA	1061	9.42	1059	9.44
	3	LO + 2TO + LA	1042	9.60	1042	9.60
	4	3TO + LA	1023	9.77	1023	9.77
	5	3LO + TA	1013	9.87	1012	9.88
	6	2LO + TO + TA	994	10.06	996	10.03
	7	2TO + LO + TA	975	10.26	975	10.26
	8	3TO + TA	956	10.46	959	10.43
A		3LO	948	10.55	949	10.54
B		2LO + TO	929	10.76	928	10.78
C		LO + 2TO	910	10.99	911	10.98
	9	2LO + 2LA	896	11.16	896	11.16
D		3TO	891	11.22	893	11.20
	10	3LO − TA	883	11.32	883	11.33
	11	LO + TO + 2LA	877	11.40	876	11.42
	12	2LO + TO − TA	864	11.57	865	11.56
	13	2TO + 2LA	858	11.65	858	11.65
	14	LO + 2TO − TA	845	11.83	846	11.82
	15	2LO + LA + TA	829	12.06	828	12.07
	16	3LO − LA	816	12.25	816	12.26
	17	LO + TO + LA + TA	810	12.34	809	12.36
	18	2LO + TO − LA	797	12.55	795	12.57
	19	2TO + LA + TA	791	12.64	792	12.63
	20	LO + 2TO − LA	778	12.85	778	12.85
E		2LO + LA	764	13.09	765	13.07
	21	3TO − LA	759	13.17	760	13.15
F		LO + TO + LA	745	13.42	745	13.42
	22	LO + TO + 2TA	743	13.46	741	13.50
G		2TO + LA	726	13.77	728	13.73
	23	2TO + 2TA	724	13.81	725	13.80
	24	LO + 3LA	712	14.04	712	14.05
	25	2LO + LA − TA	699	14.31	700	14.29
H		2LO + TA	697	14.35	696	14.36
	26	TO + 3LA	693	14.43	689	14.52
	27	LO + TO + LA − TA	680	14.70	683	14.63
I		LO + TO + TA	678	14.75	675	14.83
	28	2TO + LA − TA	661	15.13	661	15.12
J		2TO + TA	659	15.17	660	15.16

[a]Reprinted by permission from W. J. Turner and W. E. Reese (Ref. 29).

pressibility. This relation is not expected to hold precisely for the zinc-blende type of crystal. However, it may be expected that the sum of the lattice frequencies should be a constant for each wave vector $\mathbf{k}$. A comparison of the Brout sum calculated at the zone boundary with its value at the zone center ($\mathbf{k} = 0$) thus should provide an additional check on the assignment of the zone-edge phonon frequencies. Such a comparison for AlSb is made below. At $\mathbf{k} = 0$

$$\omega_l^2 + 2\omega_t^2 = 1.13 \times 10^{28} \text{ sec}^{-2} \tag{67}$$

From the frequencies LO, TO, LA, and TA of Table V, the value of the Brout sum for some $\mathbf{k}$ at or near the zone boundary is

$$LO^2 + 2TO^2 + LA^2 + 2TA^2 = 1.08 \times 10^{28} \text{ sec}^{-2} \tag{68}$$

A value of $1.11 \times 10^{28} \text{ sec}^{-2}$ is obtained when the LO and TO designations are reversed. In Eqs. (67) and (68) the factor 2 arises from the assumed degeneracy of the transverse modes.

Lattice Infrared Spectra of Diatomic Cubic Crystals

Alkali Halides. Czerny [63], Barnes and Czerny [64], Barnes [65], Hohls [66], Klier [67], Heilman [68], Abeles and Mathieu [69], and Frohlich [70] have measured the absorption and reflection spectra of a large number of alkali halides. Hass [71] has recently studied the reflection spectra of LiCl, LiBr, KF, RbF, and CsF which were not previously investigated. In addition to the absorption maximum corresponding to the transverse optical mode, a few subsidiary maxima on the short-wavelength side of the absorption and reflection spectra are also observed. However, the number of multiphonon absorption bands in the alkali halides is less than that in a III − V compound such as AlSb, discussed in the earlier section. It appears that the extent of multiple phonon interactions giving rise to absorption maxima in a crystal depends markedly on the effective ionic charge. For example, the spectra of the II − VI compounds, which are less ionic than the alkali halides or the alkaline earth halides [72], show more structure than the spectra of the latter two types of solids. The multiphonon structure is even more pronounced in III − V compounds and in homopolar crystals like diamond.

Hass [73] has studied the reflection spectrum of NaCl from

300 to 985°K in the region of the fundamental lattice absorption. The damping constant, deduced from a single dispersion formula, varied approximately with the square of the temperature. This is shown to be consistent with theoretical expectation.

The alkali halides do not show first-order Raman spectra. The second-order spectrum has recently been reviewed by Mitra [74] and thus need not be repeated here. Stekhanov and Eliashberg [75] have reported on the spectra of KBr and KBr−KCl mixed crystals. They find that in addition to a second-order spectrum, in the mixed crystals there occurs a first-order spectrum due to the lattice defects.

Table VII lists the zone-center phonon frequencies ν_t and ν_l (calculated from the Lyddane−Sachs−Teller formula [26]), the high- and low-frequency dielectric constants ϵ_∞ and ϵ_0, and the Szigeti effective ionic charge q^*. Hanlon and Lawson [76] have noticed a correlation between the values of q^* and the difference in atomic polarizability calculated from the Clausius−Mosotti formula [77]. The zone-boundary phonon frequencies of a few alkali halides have so far been determined using neutron spectrometry [78].

TABLE VII
Lattice Vibration Data on Alkali Halides[a]

Compound	ν_t,cm^{-1}	ϵ_0	ϵ_∞	ν_l,cm^{-1}	q^*
LiF	307	9.27	1.92	675	0.87
LiCl[b]	191	11.05	2.747	398	0.73
LiBr[b]	159	12.1	3.161	325	0.68
NaF	246	6.0	1.74	457	0.93
NaCl	164	5.62	2.25	259	0.74
NaBr	134	5.99	2.62	203	0.69
NaI	117	6.60	2.91	176	0.71
KF[b]	190	6.05	1.548	326	0.88
KCl	141	4.68	2.13	209	0.80
KBr	113	4.78	2.33	162	0.76
KI	98	4.94	2.69	133	0.69
RbF[b]	156	5.91	1.926	286	0.95
RbCl	118	5.0	2.19	178	0.84
RbBr	88	5.0	2.33	129	0.82
RbI	77	5.0	2.63	106	0.89

[a]Data from Ref. 25 unless otherwise stated.
[b]From Ref. 71.

TABLE VIII
Zone-Center Phonon Frequencies and the Effective Ionic Charge of Some II – VI Compounds[a]

Compound	ν_t, cm^{-1}	ϵ_0	ϵ_∞	ν_l	q^*
CdTe	140	10.6	7.13	169	0.742
CdSe	186	10.0	7.02	218	0.714
CdS	241	9.2	5.24	319	1.00
ZnTe	190	10.1	8.26	210	0.530
ZnSe	209	8.1	5.75	250	0.706
ZnS[b]	312	7.5	5.1	379	0.854
ZnO[c]	414	8.15	4.0	591	1.07

[a]Data from Ref. 79 if not stated otherwise.
[b]From Ref. 80.
[c]From Ref. 81.

II – VI Compounds. The infrared lattice spectra of CdTe, CdSe, CdS, ZnTe, ZnSe, ZnS, and ZnO have been reported [79–81] either in reflection or in absorption. Except for CdSe, CdS, and ZnO, which are hexagonal, these crystals belong to the cubic zinc-blende structure. ZnS occurs in both modifications. Table VIII gives the values of the transverse and longitudinal optical phonon frequencies at or near $\mathbf{k} = 0$, along with the value of effective ionic charge. Multiphonon structure has been observed in CdS by Balkanski and Besson [82], Deutsch [83], Mitra [84], and Marshall and Mitra [85]; in cubic ZnS by Deutsch [80]; in hexagonal ZnS by Marshall and Mitra [85]; and in ZnSe by Aven, Marple, and Segall [86]. For CdS several assignments of the zone-edge characteristic phonon frequencies are possible. Marshall and Mitra [85] have shown that Deutsch's assignment for cubic ZnS is erroneous, and have given a new assignment. They have also investigated the transmission spectrum of hexagonal ZnS, which differs slightly from the cubic form in the positions of the absorption bands. Mitra [87] has furnished an assignment for the multiphonon structure observed in ZnSe. The characteristic zone-boundary phonon frequencies of the II – VI compounds are presented in Table IX. Contrary to the alkali halides, a linear correlation has been noted by Marshall and Mitra [88] between the values of q^* and the difference between the polarizability of group VI and group II atoms, as shown in Fig. 19.

TABLE IX
Zone-Boundary Phonon Frequencies of Some II – VI Compounds

Compound	Structure	LO, cm⁻¹	TO, cm⁻¹	LA, cm⁻¹	TA, cm⁻¹
ZnSe[a]	Zinc blende	212	208	162	87
ZnS[b]	Zinc blende	339	298	155	93
				or 190	or 115
ZnS[b]	Wurtzite	346	318:297	181	92;73
ZnO[c]	Wurtzite	489	441;426	243	137;101
CdS[b]	Wurtzite	295	261;238	149	79;70

[a]Data from Ref. 87.
[b]Data from Ref. 85.
[c]Data from Mitra and Marshall (unpublished).

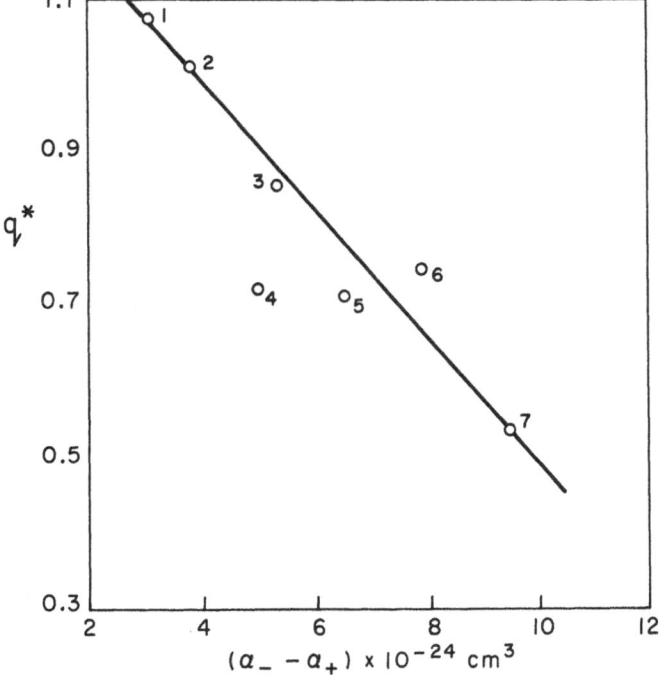

Fig. 19. q^* versus $(\alpha_- - \alpha_+)$ for the II–VI compounds: 1, ZnO; 2, CdS; 3, CdSe; 4, ZnS; 5, ZnSe; 6, CdTe; 7, ZnTe. (S. S. Mitra and R. Marshall, unpublished.)

TABLE X
Zone-Center Phonon Frequencies and the Effective Ionic Charge of Some III – V Compounds[a]

Compound	ν_t, cm^{-1}	ϵ_0	ϵ_∞	ν_l, cm^{-1}	q^*
InSb	185	17.88	15.68	197	0.42
InAs	219	15.15	12.25	243	0.56
InP	307	12.61	9.61	351	0.68
GaSb	231	15.69	14.44	240	0.33
GaAs	273	12.90	10.90	297	0.51
GaP[b]	366	10.18	8.46	402	0.58
AlSb[b]	318	12.04	10.24	345	0.53

[a]Data from Ref. 89 if not stated otherwise.
[b]From Ref. 90.

III – V Semiconductors. The restrahlen spectra of InSb, InAs, GaSb, GaAs, InP, and AlSb have recently been investigated by Hass [89] and that of GaP by Kleinman and Spitzer [90]. These crystals belong to the zinc-blende structure. The values of ν_t, ν_l, ϵ_0, ϵ_∞, and q^* are listed in Table X. A linear dependence of q^* on the difference between the atomic polarizabilities of the group III and group V atoms has been noted by Hass. A qualitative explanation of this behavior has been offered by Matossi [91].

The characteristic zone-boundary phonon frequencies of GaP [90], InSb [92], GaAs [93], and AlSb [28] have been obtained from the analyses of their transmission spectra. Newman [94] has observed absorption bands of InP in the $15.2 - 26.4\,\mu$ region. Hilsum and Rose-Innes [95] have tentatively assigned these to the multiphonon combinations. Harman, Genco, Allred, and Goering [96] have reported more structure in the $10 - 14\,\mu$ region. Hrostowski and Fuller [97] have observed structure in the lattice absorption spectrum of GaSb, which has been assigned to multiphonon combinations by Mitra [87]. The values of the characteristic zone-boundary phonon frequencies for some III – V compounds are listed in Table XI.

IV – IV Compounds. In homopolar crystals such as Ge and Si, the factor group fundamental is infrared inactive because the dipole moment is zero. However, type I diamonds do exhibit an absorption band corresponding to this mode. The type II diamonds, on the

TABLE XI
Zone-Boundary Phonon Frequencies of Some III – V Compounds[a]

Compounds	LO, cm^{-1}	TO, cm^{-1}	LA, cm^{-1}	TA, cm^{-1}
InSb	155	179	118	43
InP	318	329	150	62
GaSb	193	215	134	49
GaAs	234	258	188	70
GaP	361	378	197	115
AlSb	297	316	132	65

[a]Data from Ref. 87.

other hand, do not show this fundamental absorption. The factor group fundamental is Raman active, and both type I and type II diamonds show strong first-order Raman scattering in the neighborhood of the zone-center transverse optical mode at 1332 cm^{-1}. For homopolar cubic crystals, this also corresponds to the longitudinal optical mode. The interesting spectroscopic difference between the type I and II diamonds will not be further elaborated here. The reader is instead referred to an earlier review [98]. For Ge and Si the zone-center optical modes have been determined by Brockhouse and his co-workers [99, 100] from inelastic neutron-scattering measurements.

The multiphonon absorption in diamond has been described by Hardy and Smith [101], that in Ge by Brockhouse and Iyengar [99], and that in Si by Johnson [102]. The lattice absorption in SiC has been reported by Spitzer, Kleinman, and Frosch [103] and by Patrick and Choyke [104]. SiC occurs in both hexagonal and zinc-blende structures, but the infrared spectra of the two are slightly different. SiC is partially ionic and hence a separation of the transverse and longitudinal modes at the Brillouin zone center takes place. The relevant data on the IV – IV compounds are presented in Tables XII and XIII.

Trends in the Characteristic Phonon Frequencies of Crystals of the Zinc-Blende Type

As already pointed out, the multiple structure observed in the lattice spectra of the zinc-blende type of crystals can usually be accounted for in terms of four characteristic phonon energies. These

TABLE XII
Zone-Center Phonon Frequencies of Some Group IV Semiconductors

Substance	ν_t,cm^{-1}	ν_l,cm^{-1}	q^*
C[a]	1330	1330	0
Si[b]	510	510	0
Ge[c]	300	300	0
SiC[d]	793	970	0.94

[a]From Ref. 98.
[b]From Ref. 100.
[c]From Ref. 99.
[d]From Ref. 103.

TABLE XIII
Zone-Boundary Phonon Frequencies
of Some Group IV Semiconductors

Substance	LO,cm^{-1}	TO,cm^{-1}	LA,cm^{-1}	TA,cm^{-1}
C[a]	1161	1275	991	750
Si[b]	413	482	333	139
Ge[c]	247	280	215	65
SiC[d]	851	770	540	363

[a]From Ref. 101.
[b]From Ref. 102.
[c]From Ref. 99.
[d]From Ref. 104.

belong to the four branches at the Brillouin zone boundary: transverse acoustic, longitudinal acoustic, longitudinal optical, and the transverse optical. In certain cases, however, some doubts exist concerning the identification of the optical branch phonons as transverse and longitudinal modes. No such difficulty exists for the acoustic branches. Here the transverse frequency is smaller than the longitudinal one for all values of the propagation vector (**k**), except at the zone center, where both are zero. Keyes [105] has noted some correlation between the zone-boundary phonon energies of crystals of the zinc-blende type with the mass ratio of their constituent atoms and with their ionicity (q^*). The latter correlation enables one to as-

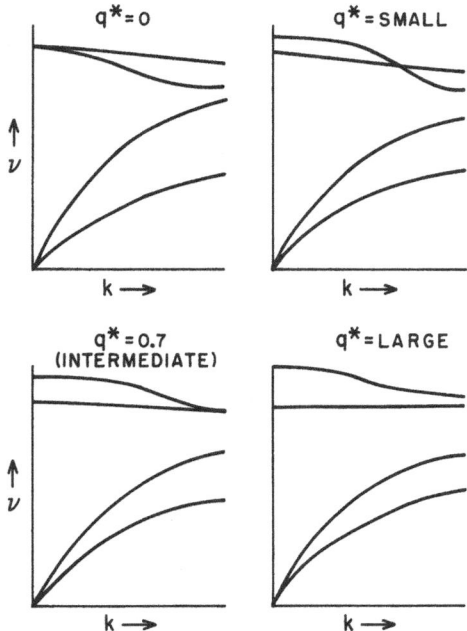

Fig. 20. The effect of ionicity on dispersion
curves.

sign the higher two optical phonon frequencies to definite longitudinal and transverse branches.

For homopolar crystals like Si ($q^* = 0$), the optical branches are degenerate at the zone center. However, the transverse optical phonons have higher energies than the longitudinal ones at or near the zone boundary. For crystals of low ionicity, the longitudinal optical mode is at a higher frequency compared to the transverse mode (Lyddane – Sachs – Teller rule) at the zone center. However, near the zone boundary, the situation is reversed, with the transverse modes having higher values than the longitudinal optical modes. This is the case with InSb, GaAs, and GaSb. In the case of GaP and AlSb, however, the LO and TO assignments of Keyes are just the reverse of those of Kleinman and Spitzer [90] and Turner and Reese [29]. The LO phonons have higher energy than the TO phonons at the zone boundary as well as at the zone center for crystals of large effective ionic charge like NaCl. Figure 20 shows the effect of ionicity on the dispersion curves as discussed above. Keyes has also predicted that the frequencies of the optical branches at the zone boundary coincide for a crystal with effective ionic charge $q^* = 0.7$ (this value of q^* we term as intermediate), and this has been noted by Mitra [87] for ZnSe.

Mitra and Marshall [88] have noted that for both II – VI and III – V compounds the q^* is a linear function of $(\nu_l - \nu_t)/\nu_t$, the zone-center optical phonon separation. This indicates that for these compounds the product of reduced mass per unit cell μ and the cell volume is a constant. They also note that the zone-center and the zone-boundary Brout sums $\sum_{i=1}^{6} \omega_i(\mathbf{k})$ are not only approximately equal but are a linear function of the reciprocal of the reduced mass. This indicates that the interatomic force constants do not change appreciably from member to member in either II – VI or III – V compounds. For the alkali halides, on the other hand, this is not true. A plot of ν_t vs. $(a/\chi\mu)^{1/2}$ yields a straight line, where χ and a are, respectively, the compressibility and lattice constant.

The Brout sum is related to χ, μ, and r_0, as was noted in Eq. (66). For zinc-blende and diamond-type crystals, the appropriate equation is

$$\sum_{i=1}^{6} \omega_i(\mathbf{k})^2 = \frac{16\sqrt{3}r_0}{\chi\mu} \tag{69}$$

Equation (69), contrary to general belief [106], is obeyed by III – V and IV – IV compounds surprisingly well. The calculated and observed values of χ in a few cases are compared in Table XIV.

The Infrared Spectrum of Gypsum, CaSO$_4$ · 2H$_2$O

The structure and the vibrational modes of gypsum have been discussed previously. In this section, we describe briefly the infrared spectra of single crystals of gypsum studied by Hass and Sutherland [11] in the 450 – 3800 cm^{-1} region. The measurements consist of transmission and reflection spectra obtained with plane-polarized radiation on three different crystal sections (010, $\bar{1}$01, and $\bar{2}$01). The Raman spectrum of the crystal has been investigated by

TABLE XIV
Calculated and Experimental Values of the Compressibility for Some Simple Diatomic Crystals[a]

| Compound | Structure | Compressibility, 10^{-12} cm² · dyne⁻¹ | |
		Calculated [Eqs. (66) and (69)]	Experimental
LiF	NaCl	1.87	1.43
NaF	NaCl	2.05	2.06
NaCl	NaCl	4.18	3.97
KF	NaCl	4.40	3.14
KBr	NaCl	7.26	6.45
RbI	NaCl	9.64	9.02
CdTe	Zinc blende	3.22	2.36
ZnS	Zinc blende	1.47	1.20
CdS	Wurtzite	2.21	1.63
InSb	Zinc blende	2.078	2.132
InAs	Zinc blende	1.750	1.727
GaSb	Zinc blende	1.716	1.855
GaAs	Zinc blende	1.338	1.324
AlSb	Zinc blende	1.751	1.695
C	Diamond	0.227	0.226
Si	Diamond	1.000	1.023
Ge	Diamond	1.172	1.330
SiC	Zinc blende	0.378	(0.473)

[a]Data from Mitra and Marshall (unpublished).

Krishnan [107], Rousset and Lochet [108], and Stekhanov [109].

Hass and Sutherland observed that for plane-polarized light incident normal to the (010) plane, the maximum dichroism usually occurred when the **E** vector made an angle (ϕ) of either 9° or 99° with the a axis. For the ($\bar{1}$01) face, the **E** vector was either parallel to the C_2 axis and gave the A_u bands, or in the (010) plane with $\phi = 33°$ and gave the B_u bands. For the ($\bar{2}$01) section the **E** vector was either parallel to the C_2 axis (A_u bands) or in the (010) plane with $\phi = 99°$ (B_u bands). The transmission and reflection spectra of an (010) section of gypsum are shown in Fig. 21. The observed frequencies, intensities, and dichroism of the absorption bands of gypsum recorded by Hass and Sutherland are listed in Table XV.

The four fundamental vibration frequencies of the free sulfate ion were reported by Kohlrausch [110] from a study of the Raman

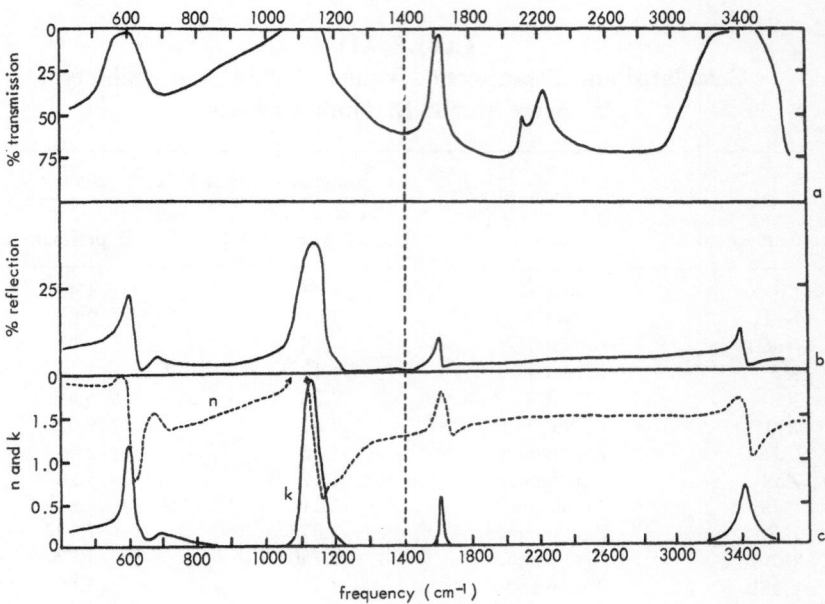

Fig. 21. Transmission (a) and reflection (b) spectra of gypsum normal to (010) face n and R deduced from middle curve are plotted in (c). (Reprinted by permission from M. Hass and G.B.B.M. Sutherland, Ref. 11.)

spectrum in solution. These along with their degeneracies and symmetry species are given below:

$$\nu_1 = 981 \text{ cm}^{-1} A(1)$$
$$\nu_2 = 451 \text{ cm}^{-1} E(2)$$
$$\nu_3 = 1104 \text{ cm}^{-1} F_2(3)$$
$$\nu_4 = 613 \text{ cm}^{-1} F_2(3)$$

Thus, in the infrared spectrum of the crystal one expects to find one band near 980 cm⁻¹ polarized parallel to the C_2 axis, which is indeed the case. The ν_1 frequency of the SO_4 ion in gypsum is found at 1000 cm⁻¹ and shows A_u character. The ν_2 mode of the sulfate ion is expected to be split into two infrared-active A_u bands near 450 cm⁻¹. However, no A_u bands were observed near 450 cm⁻¹. These bands (like the ν_1 band) may be very weak or may be outside the range of observation of Hass and Sutherland. The ν_3 mode, expected near 1104 cm⁻¹, indeed shows three components. The 1118 and 1142 cm⁻¹ bands have B_u character, while the third one at

TABLE XV
Observed Frequencies, Intensities, and Dichroism of $CaSO_4 \cdot 2H_2O$ Absorption Bands[a]

Assignment	Species	Direction[b], deg	Method[c]	Frequency, cm⁻¹	Intensity[d], cm⁻¹	Width[e], cm⁻¹
$\nu_{R'}(H_2O)$	B_u	99	R	450	—	vb
$\nu_{R''}(H_2O)$	A_u	—	R	580	—	b
$\nu_2(SO_4)$	A_u	—	R	602	30	17
$\nu_4(SO_4)$	B_u	9	R	604	30	20
$\nu_4(SO_4)$	B_u	99	R	672	35	16
$\nu_1(SO_4)$	A_u	—	R	1000	—	—
$\nu_3(SO_4)$	B_u	9	R	1118	120	32
$\nu_3(SO_4)$	A_u	—	R	1131	100	34
$\nu_3(SO_4)$	B_u	99	R	1142	130	27
$\nu_4 + \nu_4(SO_4)$	B_u	9	R	1205	—	—
$\nu_2(H_2O)$	B_u	9	R	1623	12	16
$\nu_2(H_2O)$	A_u	—	R	1685	6.5	25
$\nu_3 + \nu_1(SO_4)$	B_u	19	T	2112	—	—
$\nu_3 + \nu_3(SO_4)$	B_u	124	T	2130	—	—
$\nu_{R''} + \nu_2(H_2O)$	B_u	9	T	2198	—	—
$\nu_{R''} + \nu_2(H_2O)$	B_u	9	T	2235	—	—
$\nu_2 + \nu_2(H_2O)$	B_u	19	T	3248	—	—
$\nu_2 + \nu_2(H_2O)$	B_u	19	T	3350	—	—
$\nu_1(H_2O)$	B_u	19	R	3410	42	49
$\nu_1(H_2O)$	A_u	—	R	3430	8	b
$\nu_3(H_2O)$	B_u	—	R	3490	2	—
(H_2O)	B_u	unpolarized	T	3495	—	—
$\nu_3(H_2O)$	A_u	—	R	3537	60	67
(H_2O)	B_u	19	T	3560	—	—

[a] Taken from Hass and Sutherland (Ref. 11). [b] Direction of transition moment in (010) plane. [c] R = data derived from reflection measurements. T = data derived from transmission measurements. [d] Where numerical values are given, the number is the integrated intensity defined as $K = \int k \, d\nu$, where k is the extinction coefficient and ν is the frequency in cm⁻¹. [e] vb = very broad; b = broad.

S. S. Mitra and P. J. Gielisse

TABLE XVI
Sulfate Ion Fundamentals[a]

Solution		Gypsum	K
981 ν_1	A_g	1006	
	A_u	1000	(...)
	A_g	492	
	A_u	...	
451 ν_2	A_g	413	
	A_u	...	
	A_g	1144	
	A_u	1131	(100)
	B_g	1138	
1104 ν_3	B_u	1142	(130)
	B_g	1117	
	B_u	1118	(120)
	A_g	621	
	A_u	602	(30)
	B_g	669	
613 ν_4	B_u	672	(35)
	B_g	623.5	
	B_u	604	(30)

[a]From Hass and Sutherland (Ref. 11).

1131 cm^{-1} has A_u character in accordance with the predictions made in the section on the normal modes of gypsum. The ν_4 mode also has three components in the infrared spectrum of the crystal, the 604 cm^{-1} and 672 cm^{-1} bands belonging to the B_u species, and the 602 cm^{-1} band belonging to the A_u species. Table XVI lists 16 of the 18 fundamental vibrations assigned to the sulfate ion in gypsum. This includes also the Raman (gerade representations) assignments of Rousset and Lochet.

A comparison of the frequencies of the crystal modes with the corresponding free ion modes, reveals that all the components of ν_1 and ν_3 vibrations have higher values in the solid than in the solution. For the ν_2 and ν_4 vibrations, on the other hand, the removal of degeneracy and interactions within the unit cell have produced a distribution of frequencies around the unperturbed modes of the free ion. Since the ν_1 and ν_3 modes arise primarily from the stretching of the S—O bonds, it may be concluded that the S—O force constant is slightly higher when the sulfate ions are in the crystal

TABLE XVII
Water Fundamentals[a]

Vapor		Gypsum	K
$3657.1\ \nu_1$	A_g	3404.5	
	A_u	3430	(8)
	B_g	3402.5	
	B_u	3410	(42)
$1595\ \nu_2$	A_g	. . .	
	A_u	1685	(6.5)
	B_g	. . .	
	B_u	1623	(12)
$3755.8\ \nu_3$	A_g	3496.5	
	A_u	3537	(60)
	B_g	3498	
	B_u	3490	(2)

[a]From Hass and Sutherland (Ref. 11).

than when isolated. The degeneracy is almost completely removed in the cases of ν_2, ν_3, and ν_4 modes. The Raman frequencies corresponding to ν_2 are split by 79 cm^{-1}. Rousset and Lochet have suggested that such a large splitting may indicate that the SO_4 ion is no longer tetrahedral in the gypsum crystal but is distorted to a lower symmetry class. This may indeed be the case since two of the sulfate oxygens and the water molecules are most probably hydrogen bonded in the crystal.

The three nondegenerate fundamental vibration frequencies of the water molecule in the gaseous state are $\nu_1 = 3652$ cm^{-1}, $\nu_2 = 1595$ cm^{-1}, and $\nu_3 = 3756$ cm^{-1} [111]. In the crystal, each of them is split into four modes, two of which are infrared active (ungerade representations) and the other two Raman active (gerade). Near each of the above frequencies, one may therefore expect to find two absorption bands. One of these should be polarized parallel to the C_2 axis (A_u), with the other perpendicular to this axis (B_u). All but two of the H_2O modes in gypsum have been assigned by Hass and Sutherland. These are presented in Table XVII.

It may be noted that the hydroxyl stretching modes (ν_1 and ν_3) in the crystal have considerably lower frequencies than the corresponding free molecular modes. On the other hand, the ν_2 bending modes have higher values, which is indicative of appreciable hydrogen bonding between the H_2O molecules. The splitting of each H_2O mode in the crystal into four is due solely to the correlation field effect. Here the neighboring H_2O molecules are coupled together through hydrogen bonding to the same oxygen atom of a sulfate ion. This is in contrast to the sulfate ions themselves, which have no direct coupling. Thus the splitting of the water modes in general is considerably larger than the splitting of the corresponding modes of the sulfate ions.

Three reflection bands at 450 cm^{-1} (B_u), 580 cm^{-1} (A_u), and 1205 cm^{-1} (B_u), and a number of absorption bands observed between 2100 and 2250 cm^{-1} are yet to be assigned. Hass and Sutherland assigned the first two of these bands to rotatory lattice modes of the water molecules, in analogy with those found for ice. The rest of the bands are assigned as combination bands assuming that the factor group selection rules are obeyed, and may be regarded as tentative.

The Infrared Spectrum of Crystalline Cyclopropane

The choice of cyclopropane as an example serves two purposes. First, it is a molecular crystal with small and fairly symmetrical molecules, which makes it possible to gain a clear understanding of the influence of the crystalline field on the molecular modes. Second, the crystal structure of solid cyclopropane is not known, which makes it interesting to determine if such information can be inferred from the study of its infrared spectrum.

The cyclopropane molecule, C_3H_6, belongs to the D_{3h} point group. The 21 molecular modes of vibration are distributed among the following irreducible representations: $3A_1' + 1A_1'' + 1A_2' + 2A_2'' + 4E' + 3E''$. Barring degeneracies there are thus 14 distinct fundamental frequencies. In the infrared spectrum only the A_2'' and E' modes are permitted. In the Raman spectrum the allowed modes belong to the A_1', E', and E'' symmetry species. The modes belonging to the A_1'' and A_2' are inactive in both the spectra. The vibrational spectrum of the cyclopropane molecule has been studied by Lord and his co-workers [112] and by Mathai, Shepherd, and Welsh [113], and a satisfactory assignment of the molecular modes has been presented (see Table XVIII).

The infrared spectra of single crystals of C_3H_6 have been investigated by Brecher, Krikorian, Blanc, and Halford [114]. They

TABLE XVIII

The Fundamental Vibrations of Cyclopropane[a]

Mode	Species	Motion	Vapor, cm^{-1}		Liquid, cm^{-1}		Crystal, cm^{-1}
			Raman	IR	Raman	IR	IR
ν_1	A_1'	C—H stretch	3038	f	3027		
ν_2	A_1'	CH$_2$ deformation	1454	f	1453	1453	1454
ν_3	A_1'	Ring breathing	1188	f	1188	1191	1194
ν_4	A_1''	CH$_2$ twist	f(1133)	f	1131	1129	1133
ν_5	A_2'	CH$_2$ wag	f				1078
ν_6	A_2''	C—H stretch	f	3101		3081	3073
ν_7	A_2''	CH$_2$ rock	f	852			855
ν_8	E'	C—H stretch	3020	3025	3009	3013	3004
ν_9	E'	CH$_2$ deformation	1442	1442	1434	1432	1434
ν_{10}	E'	Ring deformation		1028	1023	1026	1027
ν_{11}	E'	CH$_2$ wag	866	866	866	865	865
ν_{12}	E''	C—H stretch	3082	f	3075		3073
ν_{13}	E''	CH$_2$ twist	1188	f	1178	1191	1200
ν_{14}	E''	CH$_2$ wag	739	f	741	741	749

[a] From Brecher et al. (Ref. 114). Bands forbidden in the vapor are denoted by the letter f.

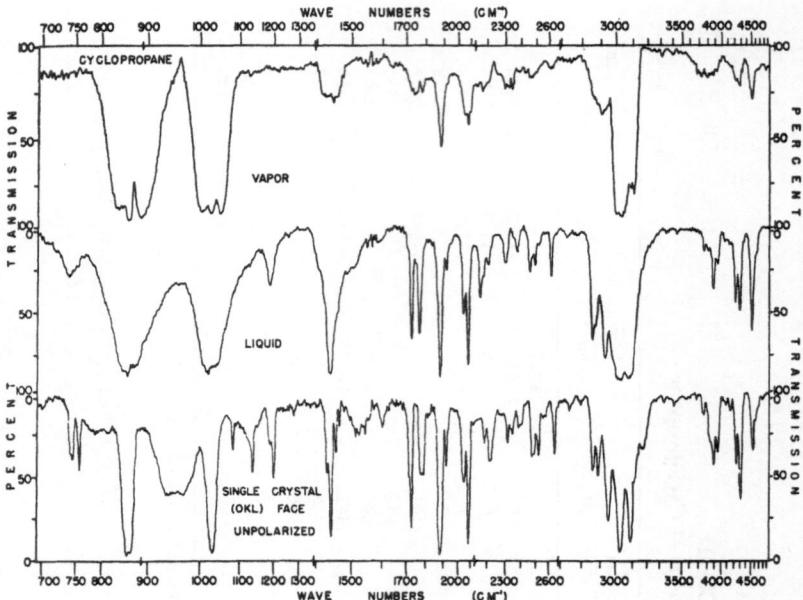

Fig. 22. Infrared spectra of cyclopropane vapor (at 25°C), liquid (at −125°C) and crystal (at −190°C), recorded with equivalent absorbing paths. (Reprinted by permission from C. Brecher *et al.*, Ref. 114).

have grown and examined two specimens with different orientations. In the absence of any x-ray crystallographic data, the relative orientations of the two samples were determined by comparing the polarized and unpolarized spectra of the two single crystals, the spectrum of a polycrystal, and also their extinctions by visible light. It was established that the two samples had (001) and (0*kl*) cross sections, respectively. Spectra were recorded at −160°C and −190°C, as was the spectrum of the liquid.

Brecher *et al.* noted that the spectra in the condensed phases were characterized by the following features when compared with the vapor phase spectrum: (i) the frequency of the absorption bands shifted very little (< 1%) from phase to phase, indicating a weak static field effect, as expected for van der Waals binding; (ii) the number of bands observed increased as one went from vapor to liquid to solid; and (iii) the PQR structure of the vapor bands was not resolved in the liquid state, which produced rather broad absorption bands, whereas the bands originating from the solid were quite sharp. Figure 22 gives such a comparison.

Since the static field shift was limited, the assignment of the internal modes of the crystal was relatively simple. However, a number of modes that are inactive in the infrared spectrum of the free molecule became active in the solid state. Some of the modes also showed correlation field splitting, which was of the same magnitude as that due to the static field at the site. A large number of bands in addition to the fundamentals were recorded for the single-crystal samples. Most of these were assigned by Brecher *et al.* as combinations of molecular modes. Satisfactory numerical agreement was obtained in most cases by using the factor group ($\mathbf{k} = 0$) fundamentals, which is not surprising in view of the limited static field effect. The observed and expected (from the polarizations of the combining frequencies) dichroism of the combination bands agreed in most cases.

A number of bands that were not assigned as either internal fundamentals or their combinations, were assigned to the combination of internal with external modes. However, since no far-infrared spectrum of the solid was recorded, the frequencies of the combining lattice modes were deduced from the combination modes. The six external degrees of freedom of a cyclopropane molecule belong to the following irreducible representations of the D_{3h} point group: $1A_2' + 1A_2'' + 1E' + 1E''$. These degrees of freedom of the isolated molecule give rise to the translatory and rotatory lattice modes in the crystal. However, since the crystal structure of the solid is not known, the exact number and nature of the lattice modes could not be enumerated with certainty. Brecher *et al.* were able to locate four primary lattice modes from the combination tones and the combining molecular mode frequencies. These are: 55 cm^{-1} (E' translation), 70 cm^{-1} (A_2' libration), 78 cm^{-1} (A_2'' translation), and 92 cm^{-1} (E'' libration). These values were ascertained from both sum and difference bands arising from combination with one or more internal modes. The internal–external combination band assignment was also consistent with the following intensity considerations: the difference (hot) bands originate from higher lattice vibrational levels and hence are weaker and much more temperature dependent that the sum bands. The complete assignment due to Brecher *et al.* is given in Table XIX.

By a judicious use of infrared spectral data, Brecher *et al.* have been successful in deducing a reasonable structure for cyclopropane crystals. This is not known from any other type of measurement. It was done by systematically testing the subgroups of the molecular

TABLE XIX
The Infrared Absorption Bands of Cyclopropane[a]

Vapor, cm⁻¹	Liquid, cm⁻¹	Crystal cm⁻¹	Polarization	Assignment
	741 m, bd	742 m	I	$\nu_{14}(E'')$
		747 m	II	
		756 m	III>>II>I	
		776[b] vvw, bd	I>II≲III	$\nu_{11} - \nu_{(Rz,Ry)}(A_1'' + A_2'' + E'')$
		798[b] w, bd	I≈II≈III	$\nu_{11} - \nu_{Rz}(E')$
841 vs P⎫ 852 pd Q'⎬ 866 vs Q⎫ 895 vs R⎭	865 vs. bd	855 vs⎫ 865 vs⎭	III>II I≈II≈III	$\nu_{7}(A_2'')$ $\nu_{11}(E')$
		937[b]s. bd⎫ 941[c]s. bd⎭	I≈II≈III	$\nu_{11} + \nu_{Rz}(E')$
		960[b]s. bd⎫ 965[c]s. bd⎭	I≳II≈III	$\nu_{11} + \nu_{(Rz,Ry)}(A_1'' + A_2'' + E'')$
1009 vs P⎫ 1028 vs Q⎬ 1051 vs R⎭	1026 vs, bd	1027 vs	I≈II≈III	$\nu_{10}(E')$
		1078 m	I	$\nu_{5}(A_2')$
		≈1096[d] vw, bd	≈I∥II∥III	$\nu_{10} + \nu_{Rz}(E')$
		≈1124[d] vw, bd	≈I∥II∥III	$\nu_{10} + \nu_{(Rz,Ry)}(A_1'' + A_2'' + E'')$

			Polarization	Assignment
≈1129 vw, sh		1133 m	I	$\nu_4(A_2'')$
1191 m, bd		1189 m	II	$\nu_{13}(E'')$ and $\nu_3(A_1')$
		1200 m		
		≈1270–75 vvw, pd	I > III	$\nu_3 + \nu_{7z}$
	1422 m, sat P	1424 m	II	$\nu_9(E')$
1432 s	1442 m, Q	1434 s	III > I	
	1455 m, sat R			
1453 w, sh		1449 m	II	$\nu_2(A_1')$
		1459 w	III	
1504 m, sh, bd		1496 vw, sat	I ≈ II ≈ III	$2\nu_{14}(E')$
		≈1509^b vvw pd	I ≈ II ≈ III	$\nu_2 + \nu_{(7x,7y)}(E')$
		≈1513^c vvw pd	II > III	$2\nu_{14}(A_1')$
		1518 m	II > III	
		1532^b w, sat	II ≈ III	$\nu_2 + \nu_{7z}(A_2'')$
		1538^c w	I > II ≈ III	$\nu_7 + \nu_{14}(E')$
		1600 w	II > III	$2\nu_7(A_1')$
1740 w		1714 m, sh	I ≈ II ≥ III	$2\nu_{11}(A_1' + E')$
	1725 s	1724 s		
1776 w	1764 m	1774 w	II ≈ III >> I	$\nu_{10} + \nu_{14}(A_1'' + A_2'' + E'')$
		1786 m	II ≈ III > I	$\nu_5 + \nu_{14}(E'')$
1881 m		1812 vw, sat	II ≈ III > I	
	1882 s	1883 s	I	$\nu_{10} + \nu_{11}(A_1' + A_2'' + E')$
		1894 s, sh		

TABLE XIX Continued

Vapor, cm⁻¹	Liquid, cm⁻¹	Crystal cm⁻¹	Polarization	Assignment
	1918 w	1923 m	I>II≥III	$\nu_5 + \nu_{11}(E')$
	2040 m	2035 m, sh	I≈II≈III	$\nu_7 + \nu_{13}(E')$
		2049 m	II>III>I	$\nu_{11} + \nu_{13}(A_1' + A_2'' + E')$
2058 m, sh P 2083 m Q 2102 m, sh R	2082 s	2083 s	II≈III	
		2091 m, sh	I	$\nu_5 + \nu_{10}(E')$
		2126 vw	II>III	$2\nu_5(A_1')$
2183 w	2169 m	2172 w	II≧III	$\nu_9 + \nu_{14}(A_1'' + A_2'' + E'')$
	2209 w, sat	2196 m	I>III	$\nu_3 + \nu_{10}(E')$
		2206 m	I	
		2217 w, sh	I≧III	$\nu_{10} + \nu_{13}(A_1'' + A_2'' + E'')$
2314 w	2304 w	2294 w	I≈II≈III	$\nu_9 + \nu_{11}(A_1' + A_2'' + E')$
		2311 w	II≈III	$\nu_2 + \nu_{11}(E')$ or
		2319 w		$\nu_4 + \nu_{13}(E')$
	2372 vw	2370 vw	I≈II≈III	$2\nu_{13}(A_1' + E')$
2456 vw	2450 w	2444 m	I>II>III	$\nu_9 + \nu_{10}(A_1' + A_2' + E')$
		2456 m	I>II>III	$\nu_2 + \nu_{10}(E')$
	2486 m	2487 m	I>I≈III	$\nu_5 + \nu_9(E')$
	2518 w	2523 w	I≈III	
		2540	II	$\nu_{10} + 2\nu_{14}(A_1' + A_2' + 2E')$
2612 w	2610 m	2608 m	I≈II≈III	$\nu_3 + \nu_9(E')$
		2726 vw	I≈II≈III	$\nu_{10} + 2\nu_{11}(A' + A_2' + 2E')$
2862 w, sh	2859 s	2855 m	II>I≈III	$2\nu_9(A_1' + E')$
2875 w, sh	2870 s	2877 m	II>I≈III	$\nu_2 + \nu_9(E')$

				Assignment
2921 m, sat	2928 s	2934 s	$I \approx II \approx III$	$\nu_9 + 2\nu_{14}\,(A_1' + A_2' + 2E')$
3003 vs, pd P 3025 vs Q 3048 vs, pd R	3013 vs	3004 vs	$I \approx II \approx III$	$\nu_8(E')$ and $\nu_1(A_1')$
3101 s, sat Q 3120 s, sat R	3081 vs	3073 vs	$II \approx III > I$	$\nu_6(A_2'')$ and $\nu_{12}(E'')$
	3178 vw, sh	3168 m, sat	$I \approx III > II$	$\nu_2 + 2\nu_{11}(A_1' + E')$
3805 vs	3814 w, sat	3824 m, sat	$I \approx II \approx III$	$\nu_6 + \nu_{14}(E')$ or $\nu_{12} + \nu_{14}(A_1' + A_2' + E')$
3892 vw	3874 m	3863 m	$I > II \approx III$	$\nu_1 + \nu_{11}(E')$ or $\nu_8 + \nu_{11}(A_1' + A_2' + E')$
	3929 w, sat	3918 m	$I > II \approx III$	$\nu_7 + \nu_{12}(E')$
4206 w, sat	4201 m	4194 m	$II > III > I$	$\nu_3 + \nu_8(E')$ and $\nu_8 + \nu_{13}(A_1'' + A_2' + E')$
4281 w	4276 m	4264 s	$I \approx II \approx III$	$\nu_6 + \nu_{13}(E')$ or $\nu_{12} + \nu_{13}(A_1' + A_2' + E')$
4522 w	4517 s	4510 m	$II \approx III \gg I$	$\nu_9 + \nu_{12}(A_1'' + A_2'' + E')$
		4560 m, sh	$II \approx III$	$\nu_2 + \nu_6(A_2'')$ and $\nu_{12} + 2\nu_{14}(A_1'' + A_2' + 2E'')$
	4621 vw	4614 w, sh	$II \approx III$	$\nu_6 + 2\nu_{14}(A_1' + A_2' + 2E')$

[a] From Brecher et al. (Ref. 114). Translatory and rotatory lattice modes are denoted ν_T and ν_R, respectively, molecular directions with which the motions are associated are included in the subscript. Symbols: vs = very strong; s = strong; m = medium; w = weak; vw = very weak; vvw = very, very weak; bd = broad; pd = poorly defined. sh = shoulder.

[b] Appears only in the spectrum taken at −160°C.

[c] Appears only in the spectrum taken at −190°C.

[d] Appears at both temperatures, but it is too poorly defined for precise measurement.

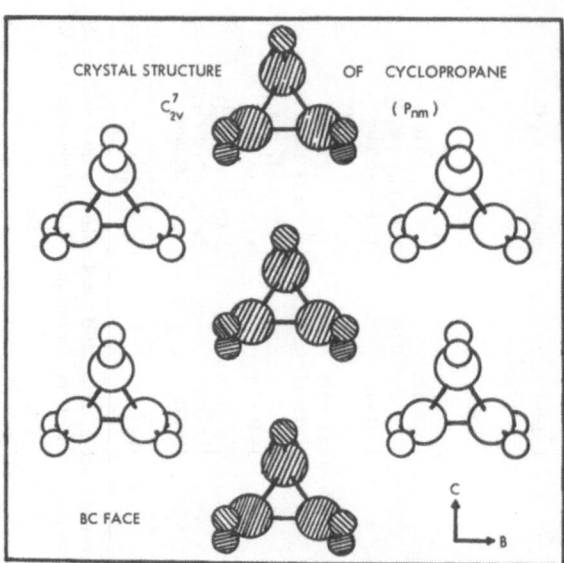

Fig. 23. Proposed crystal structure of cyclopropane, belonging to space group C_{2v}^7 (P_{nm}). The shaded molecules are not in the same plane as the unshaded ones, and are inclined oppositely. (Reprinted by permission from C. Brecher *et al.*, Ref. 114.)

point group (which is known) for selection rules which are in accord with the spectral data on the crystal. The acceptable subgroups were further examined within possible factor groups to predict the observed polarization properties of the absorption bands. It was found that only two site groups C_2 and C_s (σv) explain the appearance in the crystal spectra of the bands that are forbidden by the molecular selection rules. Now, by considering the dichroism of the absorption bands, the first of the two possibilities may be eliminated, leaving C_s (σv) as the unique site group. The occurrence of a set of three crystal fundamentals corresponding to a single molecular fundamental (e.g., ν_{14}), such that each component appears exclusively along one of the three orthogonal and nonequivalent axes, suggests that the cyclopropane crystal belongs to the orthorhombic system with only three possible isomorphous factor groups: C_{2v}, D_2, and D_{2h}. Of these only C_{2v} and D_{2h} are compatible with the $C_s(\sigma v)$ site symmetry. There are 22 (10 primitive) space groups corresponding to the factor group C_{2v} and 28 (16 primitive) space groups isomorphous with the factor group D_{2h}. A few of these are eliminated on the ground that they do not contain reflection planes as sites. This still leaves 30 possible space groups to which the cyclopropane crystal may belong.

To supplement the spectroscopic data, Brecher *et al.* now assume that in the crystal the cyclopropane molecules are packed as closely as their geometry allows. This is usually the case with most hydrocarbons. It is also noted that crystalline cyclopropane is a trifle heavier than the liquid. Since the structure of a single molecule is completely known, all the possible space groups are examined against the above criteria. Brecher *et al.* conclude that C_{2v}^7 (*Pnm*) is the most likely space group.

There are two symmetry elements in the C_s site group, whereas the C_{2v} factor group consists of four elements. Since the number of spatially nonequivalent molecules per primitive unit cell is given by the ratio of the order of the factor group to the order of the site group, there must be two molecules per primitive (Bravais) unit cell. Furthermore the C_{2v}^7 space group in an orthorhombic unit cell requires that one molecule be at the corner and the other at the body center. The crystal structure of cyclopropane as proposed by Brecher *et al.* is shown in Fig. 23.

ACKNOWLEDGMENTS

Our thanks are due to Dr. S. Nudelman and Dr. J. N. Plendl for their interest. One of us (S. S. M.) is grateful to Dr. M. R. Salk for his continuing encouragement. The writing of this article was supported by a contract (No. AF19(628)-2418) from U.S. Air Force Cambridge Research Laboratories, Office of Aerospace Research.

REFERENCES

1. Mitra, S. S., Solid State Physics: Advances in Research and Applications, Vol. 13, Academic Press, New York (1962).
2. Veddar, W., and Hornig, D. F., *Advances in Spectr.* 2:189 (1961).
3. Mathieu, J. P., *J. phys. radium* 16:219 (1955).
4. Born, M., and Von Kármán, T., *Physik. Z.* 13:297 (1912).
5. Bhagavantam, S., and Venkatarayudu, T., *Proc. Indian Acad. Sci.* A9:224 (1939); Bhagavantam, S., Ibid. A13:543 (1941).

6. Rosenthal, J. E., and Murphy, G. M., *Rev. Modern Phys.* 8:317 (1936).
7. Halford, R. S., *J. Chem. Phys.* 14:8 (1946).
8. Hornig, D. F., *J. Chem. Phys.* 16:1063 (1948).
9. Winston, H., and Halford, R. S., *J. Chem. Phys.* 17:607 (1949).
10. Mitra, S. S., *Z. Krist.* 116:149 (1961).
11. Hass, M., Dissertation, University of Michigan, 1955; Hass, M., and Sutherland, G. B. B. M., *Proc. Roy. Soc. (London)*, A 236:427 (1956).
12. Ambrose, E. J., Elliot, A., and Temple, R. B., *Proc. Roy. Soc. (London)*, A 206:192 (1951).
13. Pimentel, G. C., and McClellan, A. L., *J. Chem. Phys.* 20:270 (1952).
14. Angell, C. L., pp. 1–45 in this volume.
15. Horning, D. F., *J. Chem. Phys.* 16:1063 (1948).
16. West, W., and Edwards, R. T., *J. Chem. Phys.* 5:14 (1937).
17. Bauer, H., and Magat, M., *J. phys. radium* 9:319 (1938).
18. Bellamy, L. J., *Spectrochim. Acta* 14:193 (1959).
19. Pullin, A. D. E., *Spectrochim Acta* 13:125 (1958); *Proc. Roy. Soc. (London)*, A 255:39 (1960).
20. Buckingham, A. D., *Proc. Roy. Soc. (London)*, A 248:169 (1958); A 255:32 (1960).
21. Hexter, R. M., *J. Chem. Phys.* 25:1286 (1956).
22. Dows, D. A., *J. Chem. Phys.* 29:484 (1958).
23. Dows, D. A., *J. Chem. Phys.* 32:1342 (1963).
24. Huang, K., *Proc. Roy. Soc. (London)*, A 208:352 (1951); Born, M., and Huang, K., Dynamical Theory of Crystal Lattices, Oxford, New York, (1954), p. 82.
25. Born, M., and Huang, K., *Ibid.,* p. 396.
26. Lyddane, R. H., Sachs, R. G., and Teller, E., *Phys. Rev.* 59:673 (1941); Casselman, T. N., Mitra, S. S., and Spector, H. N., *J. Phys. Chem. Solids* (in press).
27. Haas, C., and Ketelaar, J. A. A., *Phys. Rev.* 103:564 (1956).
28. Spitzer, W. G., Kleinman, D., and Walsh, D., *Phys. Rev.* 113:127 (1959).
29. Turner, W. J., and Reese, W. E., *Phys. Rev.* 127:126 (1962).
30. Havelock, T. H., *Proc. Roy. Soc. (London)*, A 105:488 (1924); the idea of centro-frequency developed by Plendl [*Phys. Rev.* 119:1598 (1960)] is sometimes useful in characterizing the lattice vibration of a solid by one simple frequency.
31. Woods, A. D. B., Cochran, W., and Brockhouse, B. N., *Phys. Rev.* 119:980 (1960); Woods, A. D. B., Brockhouse, B. N., Cowley, R. A., and Cochran, W., *Ibid.* 131:1025 (1963).
32. Berreman, D. W., *Phys. Rev.* 130:2193 (1963).
33. Simon, I., *J. Opt. Soc. Am.* 41:336 (1951).
34. Robinson, T. S., *Proc. Phys. Soc. (London)*, B 65:910 (1952); Robinson, T. S., and Price, W. C., *Ibid.* B 66:969 (1953).
35. Schatz, P. N., Maeda, S., and Kozima, K., *J. Chem. Phys.* 38:2658 (1963).
36. Bowlden, H. J., and Wilmshurst, J. K., *J. Opt. Soc. Am.* 53:1073 (1963).
37. Fahrenfort, J., *Spectrochim. Acta* 17:698 (1961).
38. McMahon, M. O., *J. Opt. Soc. Am.* 40:376 (1960).
39. Haas, C., Dissertation, Amsterdam, 1956; Ketelaar, J. A. A., and Haas, C., *Physica* 22:1283 (1956).
40. Stierwalt, D. L., and Potter, R. F., Proceedings of the International Conference on Physics of Semiconductors, Exeter 1962 (The Institute of Physics and Physical Society, London, 1962), p. 513.

41. Born, M., and Blackman, M., *Z. Physik* 82:551 (1933); Blackman, M., *Ibid.* 86:421 (1933); *Trans. Roy. Soc. (London)*, A 226:102 (1936); Barnes, R. B., Brattain, R. R., and Seitz, F., *Phys. Rev.* 48:582 (1935); Kleinman, D. A., *Ibid.* 118, 118 (1960).

42. Lax, M., and Burstein, E., *Phys. Rev.* 97:39 (1955).

43. Birman, J. L., *Phys. Rev.* 127:1093 (1962); 131:1489 (1963); Mitra, S. S., *Physics Letters* 11:119 (1964).

44. Ziman, J. M., Electrons and Phonons, Clarendon Press, Oxford, 1960, Chapter Cochran, W., Fray, S. J., Johnson, F. A., Quarrington, J. E., and Williams, N., *J. Appl. Phys.* 32:2102 (1961).

45. Mitra, S. S., *J. Chem. Phys.* 39:3031 (1963).

46. Mitra, S. S., *Z. Krist.* 116:149 (1961).

47. Vegard, L., *Nature* 124:267 (1929); 125:14 (1930).

48. Smith, A. L., Keller, W. E., and Johnston, H. L., *Phys. Rev.* 79:728 (1950).

49. Ewing, G. E., and Pimentel, G. C., *J. Chem. Phys.* 35:925 (1961); Ewing, G. E., *Ibid.* 37:2250 (1962).

50. Dows, D. A., *Spectrochim. Acta* 13:308 (1959).

51. Gush, A. P., Hare, W. F. J., Allin, E. J., and Welsh, H. L., *Can. J. Phys.* 38:176 (1960).

52. Mitra, S. S., Solid State Physics: Advances in Research and Applications, Vol. 13, Academic Press, New York (1962), pp. 66–78; Wickersheim, K. A., *J. Chem. Phys.* 31:863 (1959); Buchanan, R. A., Kinsey, E. L., and Caspers, H. H., *Ibid.* 36:2665 (1962); Buchanan, R. A., and Caspers, H. H., *Ibid.* 38:1025 (1963).

53. Mitra, S. S., Crystallography and Crystal Perfection, Ramachandran, G. N., (ed.), Academic Press, Inc., London, 1963, pp. 347–57.

54. Johnson, F. A., and Cochran, W., Proceedings of the International Conference on Physics of Semiconductors, Exeter 1962 (The Institute of Physics and Physical Society, London, 1962), p. 498.

55. Hexter, R. M., and Dows, D. A., *J. Chem. Phys.* 25:504 (1956).

56. Mitra, S. S., Dissertation, University of Michigan, Ann Arbor, Michigan, 1957.

57. Buchanan, R. A., *J. Chem. Phys.* 31:870 (1959).

58. Snyder, R. G., Kumamoto, J., and Ibers, J. A., *J. Chem. Phys.* 33:1171 (1960).

59. Balkanski, M., and Nazarewicz, W., *J. Phys. Chem. Solids* 23:573 (1962); Dawber, P. G., and Elliott, R. J., *Proc. Phys. Soc. (London)*, 81:453 (1963); Szigeti, B., *J. Phys. Chem. Solids* 24:225 (1963).

60. Lawson, K. E., Infrared Absorption of Inorganic Substances, Reinhold, New York, 1961.

61. Szigeti, B., *Trans. Faraday Soc.* 45:155 (1949).

62. Brout, R., *Phys. Rev.* 113:43 (1959).

63. Czerny, M., *Z. Phys.* 65:600 (1930).

64. Barnes, R. B., and Czerny, M., *Z. Phys.* 72:447 (1931).

65. Barnes, R. B., *Z. Phys.* 75:723 (1932).

66. Hohls, H. W., *Ann. Physik* 29:433 (1937).

67. Klier, M., *Z. Phys.* 150:49 (1958).

68. Heilmann, G., *Z. Phys.* 152:368 (1958).

69. Abeles, F., and Mathieu, J. *Ann. Phys.* 3:5 (1958).

70. Frohlich, D., *Z. Phys.* 169:114 (1962).

71. Hass, M., *J. Phys. Chem. Solids,* 24:1159 (1963).
72. Kaiser, W., Spitzer, W. G., Kaiser, R. H., and Howarth, L. E., *Phys. Rev.* 127:1950 (1962); Krishnan, R. S., and Narayanan, P. S., *Indian J. Pure and Appl. Phys.* 1:196 (1963).
73. Hass, M., *Phys. Rev.* 117:1497 (1960).
74. Mitra, S. S., Solid State Physics: Advances in Research and Applications, Vol. 13, Academic Press, New York, (1962), pp. 57–59.
75. Stekhanov, A. I., and Eliashberg, M. B., *Optics and Spectroscopy* (USSR) 10:174 (1961); Soviet Physics–Solid State 5:2185 (1964).
76. Hanlon, J. E., and Lawson, A. W., *Phys. Rev.* 113:472 (1959).
77. Born, M., and Huang, K., Dynamical Theory of Crystal Lattices, Oxford, New York, (1954), p. 106.
78. Woods, A. D. B., Cochran, W., and Brockhouse, B. N., *Phys. Rev.* 119:980 (1960); Woods, A. D. B., Brockhouse, B. N., Cowley, R. A., and Cochran, W., *Ibid.* 131:1025 (1963); Cowley R. A., Cochran, W., Brockhouse, B. N., and Woods, A. D. B., *Ibid.* 131:1030 (1963); Smart, C., Wilkinson, G. R., Karo, A. M., and Hardy, J. R., Two Phonon Infrared Absorption in NaCl Structure Ionic Crystals, presented at the International Conference on Lattice Dynamics, Copenhagen, Denmark, August 5–9, 1963.
79. Halstead, R. E., *et al., J. Phys. Chem. Solids* 22:109 (1961).
80. Deutsch, T., Proceedings of the International Conference on Physics of Semiconductors, Exeter 1962 (The Institute of Physics and Physical Society, London, 1962), p. 505.
81. Collins, R. J., and Kleinman, D. A., *J. Phys. Chem. Solids* 11:190 (1959).
82. Balkanski, M., and Besson, J. M., *J. Appl. Phys.* 32:2292 (1961).
83. Deutsch, T., *J. Appl. Phys.* 33:751 (1962).
84. Mitra, S. S., *Physics Letters* 6:249 (1963).
85. Marshall, R., and Mitra, S. S., *Phys. Rev.* 134:A1019 (1964).
86. Aven, M., Marple, D. T. F., and Segall, B., *J. Appl. Phys.* 32:2261 (1961).
87. Mitra, S. S., *Phys. Rev.* 132:986 (1963).
88. Mitra, S. S., and Marshall, R., *J. Chem. Phys.,* in press.
89. Hass, M., and Henvis, B. W., *J. Phys. Chem. Solids* 23:1099 (1962).
90. Kleinman, D. A., and Spitzer, W. G., *Phys. Rev.* 118:110 (1960); Kleinman, D. A., Ibid. 118:118 (1960).
91. Matossi, F., *J. Phys. Chem. Solids* 24:706 (1963).
92. Fray, S. J., Johnson, F. A., and Jones, R. H., *Proc. Phys. Soc.* (*London*) 76:939 (1960).
93. Cochran, W., Fray, S. J., Johnson, F. A., Quarrington, J. E., and Williams, N., *J. Appl. Phys.* 32:2102 (1961).
94. Newman, R., *Phys. Rev.* 111:1518 (1958).
95. Hilsum, C., and Rose-Innes, A. C., Semiconducting III–V Compounds, Pergamon Press, New York 1961, p. 180.
96. Harman, T. C., Genco, J. I., Allred, W. P., and Goering, H. L., *J. Electrochem. Soc.* 105:731 (1958).
97. Hrostowski, H. J., and Fuller, C. S., *J. Phys. Chem. Solids* 4:155 (1958).
98. Mitra, S. S., Solid State Physics: Advances in Research and Applications, Vol. 13, Academic Press, New York, (1962). p. 59.
99. Brockhouse, B. N., and Iyengar, P. K., *Phys. Rev.* 111:747 (1958).

100. Brockhouse, B. N., *Phys. Rev. Letters* 2:256 (1959).
101. Hardy, J. R., and Smith, S. D., *Phil. Mag.* 6:1163 (1961).
102. Johnson, F. A., *Proc. Phys. Soc. (London)* 73:265 (1959).
103. Spitzer, W. G., Kleinman, D. A., and Frosch, C. J., *Phys. Rev.* 113:133 (1959).
104. Patrick, L., and Choyke, W. J., *Phys. Rev.* 123:813 (1961).
105. Keyes, R. W., *J. Chem. Phys.* 37:72 (1962).
106. Maradudin, A. A., Montroll, E. W., and Weiss, G. H., Theory of Lattice Dynamics in the Harmonic Approximation, Academic Press, New York, 1963, p. 115. Also see Rosenstock, H. B., *Phys. Rev.* 129:1959 (1963).
107. Krishnan, R. S., *Proc. Indian Acad. Sci.* A 22:274 (1945).
108. Rousset, A., and Lochet, R., *J. phys. radium* 6:57 (1945).
109. Stekhanov, A. I., *Doklady Acad, Nauk SSSR* 92:281 (1953).
110. Kohlrausch, K. W. F., "Ramanspektren" in: Hand - und Jahrbuch der Chemischen Physik 9:399 (1943), Leipzig: Akademische Verlags Gesellschaft.
111. Herzberg, G., Infrared and Raman Spectra of Polyatomic Molecules, D. Van Nostrand, New York, 1945, p. 281.
112. Baker, A. W., and Lord, R. C., *J. Chem. Phys.* 23:1636 (1955); Gunthard, H. H., Lord, R. C., and McCubbin, T. K., Jr., *Ibid.* 25:768 (1956).
113. Mathai, P. M., Shepherd, G. G., and Welsh, H. L., *Can. J. Phys.* 34:1448 (1956).
114. Brecher, C., Krikorian, E., Blanc, J., and Halford, R. S., *J. Chem. Phys.* 35:1097 (1961).

Infrared Correlations for Organophosphorus Compounds*

John R. Ferraro

Argonne National Laboratories
Argonne, Illinois

INTRODUCTION

There has been a tremendous increase in interest in phosphorus chemistry since the end of World War II. This is probably due to the large number of varied uses phosphorus compounds have, in flame proofing, detergents, fuel additives, metal extractants (ore processing), insecticides, nerve gases, foods, plasticizers, water treatment, and more recently in inorganic polymers and possibly in cancer therapy. Many new compounds have been synthesized and then examined by infrared spectroscopy. The number of papers reporting on studies of phosphorus compounds, particularly the organophosphorus compounds, is becoming very large, and it is becoming increasingly more difficult for the worker in this field to keep up with the literature. The purpose of this paper is to present a review of the infrared spectroscopy of organophosphorus compounds, both the neutral and acidic types. In the course of preparation a Russian review article [1] has appeared, but in it the acidic compounds are treated only superficially. This review is not intended to cover the literature on this subject completely since this

*Based on work performed under the auspices of the United States Atomic Energy Commission. Presented at the Seventh Annual Infrared Spectroscopy Institute. Canisius College, Buffalo, New York, August 19–23, 1963.

would be impossible. Some infrared work appears in papers with noninfrared titles and is therefore difficult to find. However, all of the important works will be referenced.

This review will consist of the correlations in the neutral and acidic organophosphorus compounds. Each correlation will be discussed as to the scientific basis for it, the nature of the absorption, and its present status as a correlation. In addition, tables which list the frequencies found for a particular group in typical compounds are included. In conclusion the review will present the present accepted infrared correlations for these compounds.

NOMENCLATURE

The nomenclature used in this paper will follow that recommended by the Nomenclature Committee of the American Chemical Society [2], and is illustrated as follows:

Acid	Primary Ester	Secondary Ester	Tertiary Ester

(1) (Ortho) phosphoric Acids and Phosphates*

HO\	HO\	GO\	GO\
HO—P→O	GO—P→O	GO—P→O	GO—P→O
HO/	HO/	HO/	GO/

(2) Phosphonic Acids [3] and Phosphonates

HO\	HO\	GO\	GO\
H—P→O	G—P→O	(H)G—P→O	(H)G—P→O
HO/	HO/	HO/	GO/

(3) Phosphinic Acids and Phosphinates

HO\	HO\	HO\	GO\
H—P→O	G—P→O	(H)G—P→O	(H)G—P→O
H/	H/	G/	G/

(4) Phosphonous Acids [4] and Phosphonites

H—P—OH	G—P—OH	G—P—OG	G—P—OG
|	|	|	|
OH	OH	OH	OG

*G = aryl, alkyl group or variant thereof.

(5) Phosphinous Acids and Phosphinites

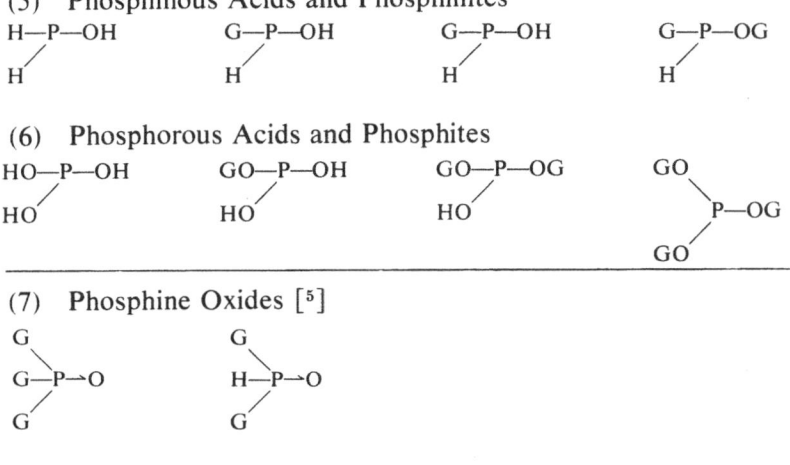

(6) Phosphorous Acids and Phosphites

(7) Phosphine Oxides [5]

(8) Phosphoramidic Acid

NEUTRAL ORGANOPHOSPHORUS COMPOUNDS

P=O Correlation

Arbuzov and his co-workers [6] studied the infrared spectra of six phosphonates of the type (alkyl—O)$_2$ PO (H). They reported the phosphoryl absorption to occur at 1250–1260 cm^{-1}. Meyrick and Thompson [7] confirmed this assignment in a study of neutral phosphonates and phosphates. Gore [8] reported an absorption at 1250 cm^{-1} in several neutral organophosphorus compounds and attributed this to the phosphoryl stretching vibration. Daasch and Smith [9, 10] investigated a large number of neutral compounds such as phosphates, phosphonates, pyrophosphates, and phosphine oxides, and suggested several empirical correlations including one for the phosphoryl absorption at 1170–1310 cm^{-1}. These results were supported by Holmstedt and Larsson [11], Bergmann et al. [12], Emeleus et al. [13], and Harvey and Mayhood [14]. Further confirmation for this absorption has come from Bellamy and Beecher [15–17], who examined 33 esters of organophosphorus acids

and assigned the phosphoryl absorption at 1250–1300 cm⁻¹.
Additional recent confirmation for this absorption has come from
Maarsen et al. [18], Gerding and Maarsen [19], Daasch [20], Houalla
and Wolf [21, 22], Burger [23, 24], Ketelaar and Gersmann [25],
Lagowski [26], Griffiths and Burg [27], Schmutzler [28], and Thomas
[29, 30]. Thomas, who has investigated a considerable number
(over 400) of neutral organophosphorus compounds, considers the
phosphoryl absorption to occur in the range 1160–1350 cm⁻¹.
The phosphoryl absorption in neutral compounds of the type
$R_2 PO (CH_2)_n POR_2$ has also recently been reported [31, 32]. The
absorption in these compounds is found in the region 1160–1266
cm⁻¹.

There appears to be little doubt as to the validity of this correlation. However, it must be realized that the range of the frequency can be quite large, because of its sensitivity to electronegative substituents. Daasch and Smith [7, 8] observed that the
position of the phosphoryl absorption appeared to be dependent on
the electronegativity of the substituent group. Bell et al. [33]
found the relationship between the phosphoryl frequency and the
electronegativity of the organo group to follow the equation

$$\lambda(\mu) = \frac{39.96 - \Sigma X}{3.995}$$

where ΣX is the sum of the phosphoryl absorption shift constants
of the substituents. Burger [23, 24] found a similar correlation between the phosphoryl absorption in neutral phosphates, phosphonates, phosphinates, and phosphine oxides, and the electronegativity
of the organic group. Table I lists the phosphoryl frequencies for a
number of neutral organophosphorus compounds.

The phosphoryl absorption in the neutral organophosphorus
compounds is very strong, and sometimes appears as a doublet
in the liquid spectrum. Mortimer [34] has proposed that the doublet
is due to the presence of more than one rotational isomer.

The phosphoryl frequency is subject to shifts of 50 cm⁻¹ or
more toward lower frequency, when the P=O becomes hydrogen
bonded or when it forms a metallic bond with metals. In addition,
since the phosphoryl dipole is exposed in neutral compounds, it
is subject to solvent effects. Several infrared studies of these effects
have been made [35–46].

TABLE I
P=O Stretching Frequencies in Several
Organophosphorus Compounds

Compound	$\nu_{P=O}$, cm^{-1}	Reference
$(CH_3O)_2 PO (F)$	1305	9
$(CH_3O)_3 PO$	1290, 1275	34
$(CH_3)_3 PO$	1176	9
$(C_2H_5O)_3 PO$	1270, 1260	34
$(C_2H_5O)_2 PO (OC_4H_9)$	1266	23, 24
$(C_4H_9O)_3 PO$	1269	23, 24
$(CH_3) (n\text{-}C_4H_9)_2 PO$	1189	23, 24
$(C_4H_9)_3 PO$	1157	23, 24
$(C_2H_5O)_2 (CH_3) PO$	1235	23, 24
$(C_6H_5)_3 PO$	1193	23, 24

P=S Correlation

The absorption associated with the P=S stretching vibration is generally weak, and has not been found to be too useful. Daasch and Smith [9] and Gore [8] studying thiophosphates found absorption of variable intensity in the region of 600–750 cm^{-1}. Bellamy and Beecher [15] found absorption in the 750 cm^{-1} region. McIvor [47, 48] found the absorption at 780–840 cm^{-1}. Thomas lists the frequency at 713–835 cm^{-1} [30].

The band appears to behave like the P=O band with respect to the effects of electronegativity groups, although the effect is perhaps smaller because of the less ionic character of the P=S bond [30a]. Thomas [29] reported the following absorptions:

$R_1 R_2 R_3 P=S$	770 and 760 cm^{-1}
$(R_1O) R_2 R_3 P=S$	770–790 cm^{-1}
$(R_1O) (R_2O) R_3 P=S$	775–805 cm^{-1}
$(R_1O) (R_2O) (R_3O) P=S$	800–845 cm^{-1}

This is in the same order as the frequency shifts for corresponding oxygen compounds. In some liquids the P=S absorption is seen as a doublet [30a], similar to the P=O absorption. This splitting may be due to the presence of rotational isomers.

The absorption of a P—S—bond should be at lower frequency, and McIvor [47, 48] has found a variable absorption at 510–575

TABLE II
P=S Stretching Frequencies in Several
Organophosphorus Compounds

Compound	$\nu_{P=S}$, cm^{-1}	Reference
$CH_3 PSCl_2$	664	49
$C_2H_5 PSCl_2$	680, 665, 640	9
$(\varphi)_2 PSCl$	735	28
$(\varphi)_2 PSF$	738	28
$(CH_3)_2 PSF$	756	28
$i\text{-}C_8H_{15} PSF_2$	~680	28
$ClCH_2PSF_2$	659	28
$(CH_3)_3 P=S$	571	30a
$(CH_3O)_3 P=S$	603, 620	30a

cm^{-1}. Thomas [30] reports a P—S—(C or H) frequency at 510–620 cm^{-1}. Here again, we are in a region in which organophosphorus compounds not containing sulfur absorb, and so one must use this correlation with caution.

The vibrations involving phosphorus and sulfur have been reported to be strong in the Raman spectra, analogous to the carbon–sulfur vibrations. Therefore, it might be more profitable to look in the Raman for this vibration.

Table II lists the P=S frequencies for several neutral organophosphorus compounds.

NEUTRAL AND ACIDIC COMPOUNDS

P—H Correlation

The P—H stretching vibration appears to be well established. Herzberg [50] reported the P—H vibrations in gaseous phosphine as occurring at 2327 and 2421 cm^{-1}, while Jost and Anderson [51] reported the vibration at 2306 cm^{-1} for liquid phosphine. Arbuzov [6] reported the vibration in some phosphonates to be centered at 2430 cm^{-1}, and Meyrick and Thompson [7] reported it to occur at 2435 cm^{-1}, for similar compounds. Daasch and Smith [9, 52] assigned the frequency at 2350–2440 cm^{-1} in several phosphonates and phosphinates as the P—H vibration. Bellamy and Beecher [15, 16], Corbridge and Lowe [53], McIvor et al. [47], Beachell and Katlafsky [54], Thomas [29, 30], and Houalla and Wolf [21] all have studied compounds having a PH bond and reported absorption in the same general region. Thomas [29] reported the PH absorption in phosphonates at 2400–2450 cm^{-1} and in phosphinates at 2325–2350 cm^{-1}.

The absorption is generally of moderate intensity, and is found to be shifted on deuteration (e.g., 1750 cm^{-1} from 2381 cm^{-1} [9]). Studies of the integrated infrared intensities of the PH vibration have been made [21]. The band appears to have a low capacity for hydrogen bonding, although Miller *et. al.* [55] cite evidence for the intermolecular association of hydrogen phosphine oxides involving hydrogen bonding of the type

Although Bellamy and Beecher [15, 16] and Daasch and Smith [9] have stated that no P—H bending vibrations have been found, Thomas [30] lists the deformation P—H vibrations at 965–1150 cm^{-1}. Table III lists the P—H stretching frequencies in several organophosphorus compounds.

P—OC Correlation

Daasch and Smith [9] proposed the region of 1030–1090 cm^{-1} as the P—OC absorption. Bellamy and Beecher [15, 16] found absorption at 1020–1053 cm^{-1} in several organophosphorus compounds, and attributed this to the P—OC vibration. Other workers [12, 53] have found similar absorption in the region 995–1000 cm^{-1}. There has been confusion as to whether this absorption is due to the (P—O)—alkyl stretch or the P—(O—alkyl) stretch, Thomas [29, 30] lists the limits of the (P—O)—alkyl absorption at 950–1055 cm^{-1}, and the P—(O—alkyl) vibration at 1170–1190 cm^{-1}. Ferraro *et al.* [56] find absorption at 1030 cm^{-1} in alkyl phosphonates. Mortimer [34] assigns the 1045 cm^{-1} band in $(CH_3O)_3$ PO and $(C_2H_5O)_3$ PO to the P—(OC) stretching vibration, the band at about 800–850 cm^{-1} to the (P—O)—C asymmetric stretching vibration, and the band at about 750 cm^{-1} to (P—O)—C symmetric stretching vibration. Nyquist [57] has found absorption in compounds containing the P—O—alkyl link at 1030 cm^{-1} and

TABLE III
P—H Stretching Frequencies in Several Organophosphorus Compounds

Compound	ν_{P-H}, cm^{-1}	Reference
$(C_4H_9O)_2\,P\overset{\displaystyle O}{\underset{\displaystyle H}{\lessgtr}}$	2410	9, 52
(C_2H_5), φ — P $\overset{\displaystyle O}{\underset{\displaystyle H}{}}$	2350	9, 52
$(CH_3O)_2\,P\overset{\displaystyle O}{\underset{\displaystyle H}{\lessgtr}}$	2436	6
$(CH_3)_2\,P\overset{\displaystyle O}{\underset{\displaystyle H}{\lessgtr}}$	2432	7
$(\varphi CH_2)_2\,P\overset{\displaystyle O}{\underset{\displaystyle H}{\lessgtr}}$	2410	15, 16
$(C_2H_5)_2\,P\overset{\displaystyle O}{\underset{\displaystyle H}{\lessgtr}}$	2420	47

780–850 cm^{-1}. Ketelaar and Gersmann [25] assign the (P—O)—alkyl asymmetric vibration at 950–980 cm^{-1}, the (P—O)—alkyl symmetric vibration at 750 cm^{-1}, and the (alkyl—O)—P stretching vibration at 1030 cm^{-1}. It is therefore uncertain which vibration is due to which mode. However, the band at ~1030 cm^{-1} is very intense, and perhaps is the most intense band in the spectra of alkyl organophosphorus compounds containing the P—O—alkyl link, and therefore, is very useful. Table IV lists the P—OC stretching frequencies for several organophosphorus compounds.

TABLE IV
POC Stretching Frequencies in Several
Organophosphorus Compounds

Compound	ν_{POC}, cm^{-1}	Reference
$(C_2H_5O)_2 CH_3 PO$	1041	25
$(C_2H_5O)_2 PO\,(OH)$	1030	25
$(CH_3O)_3 PO$	1045	34
$(C_2H_5O)_3 PO$	1030	34
$(C_2H_5O)\,(\varphi O)_2 PO$	1035	15, 16
$(C_4H_9O)_3 PO$	1030	15, 16
$(C_4H_9O)_3 P$	1030	9
$(CH_3O)_2 POF$	1052	9
$(C_4H_9O)\,(C_4H_9)\,PO\,(OH)$	1035	56
(2-ethylhexyl—O) (2-ethylhexyl) PO (OH)	1040	56

(P—O)—Aryl and P—(O—Aryl) Correlations

Bellamy and Beecher [15, 16] and Corbridge [53] found absorption for organophosphorus compounds containing the P—O—aryl group in the region of 1200–1250 cm^{-1}. Earlier, Daasch and Smith [9] found absorption at 875–950 cm^{-1}. Because of subsequent work it appears that the higher-frequency absorption is connected with the asymmetric stretching vibration of the oxygen to phenyl, while the lower-frequency absorption is connected with the (P—O) —aryl stretching vibration. Nyquist [57] has found that the (P—O)— aryl frequency is at slightly higher positions in pentavalent phosphorus compounds than in the trivalent compounds. Chapman and Harper [58] have confirmed this and report the frequency for trivalent compounds to be at 855–875 cm^{-1}. The (P—O)—aryl frequency in acidic organophosphorus compounds has been found to shift to lower frequency upon salt formation [59]. The reason for this is not completely understood.

The symmetrical stretching vibration for (P—O)—aryl has been reported by Ketelaar and Gersmann [25] to be at 743–787 cm^{-1}, and by Mortimer to be at 774 cm^{-1}. Mortimer [34] also reports a φO—P bending vibration at 470 cm^{-1}. In the KBr region compounds having a φOP group have absorption in the 600 cm^{-1} region [60]. This band appears to shift slightly depending on the ring substituents of the phenyl groups.

The peaks which appear to be studied the most thoroughly are

TABLE V
P—O—Aryl Stretching Frequencies in Several
Organophosphorus Compounds

Compound	$\nu_{(P-O)-aryl}$ asymmetric stretch, cm^{-1}	Reference
$(\varphi O)_3 PO$	960	34
$(\varphi O)_2 PO (OH)$	952	15, 16
$(p\text{-}CH_3\text{—}\varphi O)_2 PO (OH)$	990	61
$(o\text{-}CH_3\text{—}\varphi O)_2 PO (OH)$	980	61
$(p\text{-isopropyl—}\varphi O)_2 PO (OH)$	966	61
$(C_6H_5CH_2O)_2 PO (OH)$	985	15, 16
$(p\text{-}NO_2\varphi O)_2 PO (OH)$	980	15, 16
$(o\text{-}CH_3O\varphi O)_2 PO (OH)$	975	15, 16
	$\nu_{P-(O-aryl)}$ asymmetric stretch, cm^{-1}	
$(\varphi O)_3 PO$	1187, 1177	34
$(\varphi O)_2 PO (OH)$	1165	61
$(p\text{-}CH_3\varphi O)_2 PO (OH)$	1188	61
$(o\text{-}CH_3\varphi O)_2 PO (OH)$	1178	61
$(p\text{-isopropyl—}\varphi O)_2 PO (OH)$	1180	61

the (P—O)—aryl asymmetric stretching vibration at 914–994 cm^{-1} and the P—(O—aryl) asymmetric stretching vibration at 1160–1240 cm^{-1}.

The former vibration is very strong in intensity and broad. The latter peak is sharp and of medium intensity, but it appears in a region in which the strong P=O absorption also occurs, which may thus obscure the results. Further work is necessary in the case of the symmetrical stretches and bending vibrations of these groups of atoms. Table V lists frequency bands found in several typical compounds for the (P—O)—aryl and P—(O—aryl) asymmetric stretching vibrations.

P—C (Alkyl) Correlation

There is some question as to whether a correlation for this group exists. Daasch and Smith [10] proposed the strong absorption in $(CH_3)_3$ P=O at 750 cm^{-1} for the P—C (alkyl) asymmetric stretch, and a weak band (Raman) at 671 cm^{-1} for the symmetric stretch. These results were verified by Hooge and Christen [30a]. Halmann [62] investigated $(CH_3)_3$ P and related compounds, and proposed the absorption at 700 cm^{-1} as the P—C asymmetric stretch and the band at about 650–671 cm^{-1} as the symmetric P—C stretch. Deacon and Jones [63] assign the medium to strong absorption in some tetramethylphosphonium salts [e.g., $(CH_3)_4$ PHgI$_4$] at 771–780 cm^{-1} as the P—C asymmetric stretching vibration. They also report the P—C asymmetric vibration in $(CH_3)_3$ P at 704 and 714 cm^{-1}. Kaesz [64] found absorption in $(Et)_3$ P and $(vinyl)_3$ P at 655–670 cm^{-1}. However, many compounds with a P—C (alkyl) bond fail to show absorption in this region, and further work is necessary to substantiate this correlation.

Little is known of the P—C bending vibration, although Daasch and Smith [10] report it in the Raman spectra of $(CH_3)_3$P=O at about 311 and 256 cm^{-1} for the asymmetric and symmetric modes respectively.

It appears that more research is necessary with compounds containing the P—C (alkyl) bond in order to ascertain the validity of these correlations.

P—C (Aryl) Correlation

In a number of compounds containing the phenyl—P link, absorption was found at 1435–1450 cm^{-1} [9]. Another absorption was found at about 1000 cm^{-1}. Absorption in the 1000 cm^{-1} region is also found for compounds containing the P—O—phenyl group, so that this absorption is not too helpful. The higher frequency band is much more useful. The band is usually sharp and of moderate to strong intensity, and is in a region in which these compounds are nonabsorbing.

Another band which appears to be present in phenyl—P compounds (neutral and acidic) is found in the KBr region at about 530 cm^{-1} [60]. This band is usually strong in intensity and may be the strongest band in the KBr region for these compounds. Although compounds with P—O—phenyl groups also show absorption in this general region, the 530 cm^{-1} band is confirmatory for a phenyl—P link when used with the 1440 cm^{-1} peak. Table VI lists the vibration associated with the φ—P bond in a number of organophosphorus compounds.

TABLE VI
Vibrations Associated with the φ—P Bond in
Organophosphorus Compounds

Compound	$\nu_{\varphi-P}$, cm^{-1}	Reference
$\varphi_3 P$	1435	9
$(C_6H_5CH_2O)_2 C_6H_5 P{=}O$	1450	9
$\varphi_3 P{=}O$	1440	9
$(C_2H_5O)_2 C_6H_5 P{=}O$	1450	9
$C_6H_5PO (OH)_2$	1442	9
$(C_6H_5O) PO (OH)_2$	1440	9
$(n\text{-}C_4H_9O) C_6H_5PO (OH)$	1440, 530	56, 60
(Ethylhexyl—O) $C_6H_5PO (OH)$	1440, 530	56, 60
(tridecyl—O) $C_6H_5PO (OH)$	1440, 530	56, 60
$(C_6H_5CH_2O) C_6H_5PO (OH)$	1450, 530, 515	56, 60
$(C_6H_5O) C_6H_5PO (OH)$	1445, 535	56, 60

P—Halogen Correlations

The P—F and the P—Cl stretching absorptions in several organophosphorus compounds have been investigated. However, the vibration is apparently affected by the oxidation state of the phosphorus atom. In addition, the presence of more than one halogen atom causes the vibration to become either asymmetrical or symmetrical, and two peaks are observed separated from each other. As a consequence, the vibration will cover a large range. On the other hand, very little is known of the P—Br and P—I absorptions, since very few investigations have been carried out on compounds having these links.

P—F Correlation. The P—F stretching absorption in organophosphorus compounds has been investigated. Daasch and Smith [9] reported the absorption to occur at 860–900 cm^{-1} in monofluorides, and at 740 and 800 cm^{-1} in difluorides. Corbridge and Lowe [53] found the absorption at 720–835 cm^{-1} in some phosphorofluoridates (M_2PO_3F). Bellamy and Beecher [15] assigned the 853 cm^{-1} band to the P—F vibration in $[(CH_3)_2N]_2POF$. Thomas [29] reported that for pentavalent phosphorus compounds the absorption is found at 805–890 cm^{-1}, and for trivalent compounds at 760–769 cm^{-1}. Recently Schmutzler [28] reported absorption in the 884–930 cm^{-1} region for $RPOF_2$ compounds, and in the 805–835 cm^{-1} region for R_2POF compounds.

TABLE VII
P—F Stretching Frequencies in Several
Organophosphorus Compounds

Compound	ν_{P-F}, cm^{-1}	Reference
$[(CH_3)_2 N]_2 POF$	833	15
$(C_2H_5)_2 NPF_2$	740, 800	9
$(CH_3O)_2 POF$	860	9
$(C_2H_5O)_2 POF$	880	9
CH_3POF_2	930, 887	28
$C_2H_5POF_2$	906, 880	28
⬡POF_2	896, 870	28
φPOF_2	905, 875	28
$n\text{-}C_4H_9POF_2$	900, 885	28
$(CH_3)_2 POF$	808	28
$(n\text{-}C_4H_9)_2 POF$	819	28
$\varphi_2 POF$	835	28

The absorption is of strong intensity, and except for the fact that it covers a large range, appears to be a good correlation. Table VII lists the vibrations associated with the P—F bond. To date no reported assignments have been made for the rocking and deformation vibrations involving P—F atoms.

P—Cl Correlation. The P—Cl absorption would be expected to be at lower frequency from the P—F absorption because of the heavier mass of the chlorine atom. It is found in the KBr region of the infrared spectrum. Daasch and Smith [9] studied several organophosphorus compounds containing two chlorine atoms and found absorption in the 430–585 cm^{-1} region. Corbridge [53] proposed a P—Cl stretching absorption at 440–580 cm^{-1}, while Nyquist [57] reported it at 430–585 cm^{-1}. A higher range was reported by McIvor [47] at 500–570 cm^{-1}. Gerding [19] reported the absorption at 549 cm^{-1} for the asymmetric and at 496 cm^{-1} for the symmetric stretching vibration. Thomas [30] reports the P—Cl absorption at 435–587 cm^{-1} in 44 compounds.

The absorption is of strong intensity, but like the P—F vibration covers a large range. In addition, it appears in the KBr region, where organic groups also absorb, and this limits its value. Table VIII lists the P—Cl frequencies in several organophosphorus compounds.

TABLE VIII
P—Cl Stretching Frequencies in Several
Organophosphorus Compounds

Compound	ν_{P-Cl}, cm^{-1}	Reference
$\varphi_2 PCl$	521	9
φPCl_2	500	9
$C_2H_5PCl_2$	502, 485	9
$\varphi POCl_2$	572, 485	9
$\varphi_2 POCl$	521	9
$CH_3 POCl_2$	544, 494	47
$C_2H_5 POCl_2$	568, 540, 500	47
$CH_3 POCl_2$	549 asymmetrical	19
	496 symmetrical	
$CH_3 OPOCl_2$	607	65
$CH_3 PCl_2$	481	65
$CH_3 OPCl_2$	458	65

Since very little has been done in the study of organophos-phorus beyond the KBr region, very little is known regarding the rocking and deformation P—Cl vibrations. Gerding [19] reports the rocking of PCl_2 atoms in CH_3POCl_2 to occur at 357 cm^{-1} and the PCl_2 deformation to occur at 335 cm^{-1}.

P—Br Correlation. This absorption should be found beyond the KBr region because of the heavier mass of the bromine atom. This region has not been extensively studied with organophosphorus compounds, and therefore little information exists regarding the P—Br stretching or bending vibrations. Baudler [66] found the Raman absorption in organophosphorus compounds at 370 and 410 cm^{-1}. Corbridge [53] proposed the vibration at 440–485 cm^{-1}. Rice et al., [67] found absorption for phosphoronitrilic halides containing bromine in the 435–500 cm^{-1} region. Actually, an insufficient number of compounds have been investigated by in-frared to substantiate the position of the P—Br stretching vibration. Table IX lists the P—Br frequencies found in several phosphorus compounds.

P—O—P Correlation

This absorption occurs in a region where most organophos-phorus compounds have other absorptions, and therefore the

TABLE IX
P—Br Stretching Frequencies in Several
Organophosphorus Compounds

Compound	ν_{P-Br}, cm^{-1}	Reference
PBr$_3$	400, 380	50
POBr$_3$	488, 340	50
PSBr$_3$	438, 298	68
(φO) PBr$_2$ (φO)$_2$ PBr	} 370–410 (Raman)	66

correlation may not be too useful. Bergmann [12] assigned a band at 930–970 cm^{-1} to the P—O—P stretching vibration, and Harvey and Mayhood [14] list the absorption at 910–950 cm^{-1}. McIvor et al., [47] have substantiated this, and list this vibration at 910–970 cm^{-1}. Corbridge and Lowe [53] list the frequency at 850–940 cm^{-1} in pyrophosphate salts. Thomas assigns the frequency at 900–980 cm^{-1} [30]. In most cases, the peak appears to be strong. However, more work is necessary to substantiate whether absorption in this region is due to the stretching P—O—P link, and to determine where the P—O—P bending vibration occurs. Table X lists the vibrations associated with the POP group.

P=N, P—N, P—NH$_2$ (R) Correlations

Absorption in the 1200–1300 cm^{-1} region attributed to the P=N stretching vibration has been found by Daasch [69] in phosphonitrilic halide compounds. Absorption in the 680–750 cm^{-1} region has been attributed to the P—N stretching vibration by Holmstedt and Larsson [11]. Corbridge [53] found absorption near 700 cm^{-1} in several phosphorus – nitrogen compounds

$$
\begin{array}{ccc}
\text{MO} & & \text{O} \\
& \diagdown \quad \diagup & \\
& \text{P} & \\
& \diagup \quad \diagdown & \\
\text{HO} & & \text{NH}_2
\end{array}
$$

It appears that additional work is necessary in this area to elucidate the certainty of these correlations.

A recent publication by Nyquist [70] has made assignments

TABLE X
Vibrations Associated with the POP Group in Organophosphorus Compounds

Compound	ν_{POP}, cm^{-1}	Reference
$(C_2H_5O)_2$ P—O—P—$(O—C_2H_5)_2$ (with O double-bonded to each P)	950	9
$(C_4H_9O)_2$ P—O—P—$(O—C_4H_9)_2$ (with O double-bonded to each P)	950	9
$(CH_3O)_2$ P—O—P—$(O—CH_3)_2$ (with O double-bonded to each P)	970	47
$(n\text{-}C_3H_7O)_2$ P—O—P—$(O—n\text{-}C_3H_7)_2$ (with O double-bonded to each P)	958	47
$(i\text{-}C_3H_7O)_2$ P—O—P—$(O—i\text{-}C_3H_7)_2$ (with O double-bonded to each P)	982	47

for the N—H stretching and bending vibrations in phosphoramides

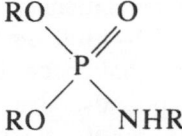

The N—H stretching vibration is located at 3400–3448 cm^{-1}, in the *cis* compounds and at 3385–3415 cm^{-1} in the *trans* compounds [70]. The N—H bending vibration, although probably coupled with other modes of vibration, is at 1372–1408 cm^{-1} [70]. Bellamy [16] reported the PNH$_2$ stretching vibration at 3125–3330 cm^{-1}, and the bending vibration at 1550–1570 cm^{-1} in solid phosphoramidates.

ACIDIC ORGANOPHOSPHORUS COMPOUNDS

P=O Correlation

The phosphoryl absorption in acidic organophosphorus compounds is affected by the intermolecular hydrogen bonding existing in these molecules. Therefore, the absorption will be found 50–80 cm^{-1} toward lower frequency.

Daasch and Smith [9] studied several phosphinic and phosphonic acids and observed the phosphoryl frequency to shift toward lower frequencies by 50–80 cm^{-1} and to broaden. Bellamy and Beecher [16] investigated acids of the type $(GO)_2PO(OH)$ and proposed that the phosphoryl absorption occurred at 1175–1250 cm^{-1}. Recent confirmation of this absorption in acids of the type $(GO)_2PO(OH)$ and $(GO)G'PO(OH)$ has been made by Ferraro et al. [56, 61]. The phosphoryl absorption in mono- and polymethylene diphosphonic acids of the type $(HO)_2OP(CH_2)_nPO(OH)_2$ has recently been reported [71] and a frequency of 1170–1220 cm^{-1} recorded. Other recent infrared studies with acidic organophosphorus compounds have been made by Thomas et al. [29, 30, 72], Hadži and Novak [73], and Winand and Dreze [74]. Thomas [30] has proposed the absorption to occur at 1140–1225 cm^{-1} in a study of 33 compounds.

The phosphoryl absorption in the acidic compounds is a very intense band and sometimes is the most intense band in the spectrum. It appears as a broad band, because of the hydrogen bonding that occurs. The acids of the types $(GO)_2PO(OH)$ [75], $(GO)G'PO(OH)$ [75], and $(G)_2PO(OH)$ [76] have been shown to be predominantly dimeric, while acids of the type $G(H)PO(OH)$ are primarily trimeric [76]. Higher degrees of aggregation have been found for acids of the types $(GO)PO(OH)_2$ and $GPO(OH)_2$ [77]. The intermolecular hydrogen bonds in these acids are extremely strong, and only when the acids are dissolved in strong basic solvents such as alcohols, acetic acid, or acetone at normal concentrations, will monomerization occur. In nonbasic solvents like CCl_4 extreme dilutions ($^1/_{30,000}$ M) and long path lengths (10 cm) are necessary to show the monomeric hydroxyl absorption [72, 78, 79]. It has been postulated that the dimeric acids form an eight-membered ring analogous to the carboxylic acids. As such the P=O dipole is already bonded, and recent infrared studies in nonbasic solvents [46] show it to be only slightly affected. The dibasic acids are more complex and little work on this type has been done to date [74, 80].

The effect of the electronegative substituents on the phosphoryl frequency of the acid phosphates $(GO)_2PO(OH)$ and the hydrogen phosphonates $(GO)G'PO(OH)$ is illustrated in Tables XI and XII. In the acid phosphates an increase in the electronegativity of

TABLE XI
Position of the Phosphoryl Absorption *vs* Electronegativity of Organo Groups in Several Acid Organophosphonates

Acid phosphonate (GO) GP (OH)		Electronegativity by Bell's method [33]	$\nu_{P=O}$, cm^{-1}
2-ethylhexyl	chloromethyl	8.3	1245
n-octyl	chloromethyl	↓	1240
phenyl	phenyl	7.9	1215
benzyl	phenyl	↑	1215
tridecyl	phenyl		1215
2-ethylhexyl	phenyl	7.7	1215
n-butyl	phenyl	↓	1215
n-octyl	phenyl		1210
2-ethylhexyl	2-ethylhexyl	7.3	1200
n-hexyl	n-hexyl		1200
n-butyl	n-butyl	↓	1197
n-decyl	n-decyl		

the substituents decreases the phosphoryl frequency. In the hydrogen phosphonates the reverse effect occurs [56, 61, 81].

The effect of metal complexing on the phosphoryl absorption of the acid phosphates is illustrated by the uranyl complexes [82, 83] of these acids, and other metal complexes [83]. The phosphoryl absorption in the acid changes on complexing, although it is not as simple and as predictable as the effect of hydrogen bonding. However, the formation of the metal complexes aids in identifying the position of the phosphoryl absorption in the acids.

P—OH Correlation

It would be expected that the hydroxyl stretching vibration would be at lower frequencies than vibrations of systems which are not hydrogen bonded, since the —OH group is involved in hydrogen bonding. Early workers [9, 15, 16] pointed out that in the dialkyl phosphates absorption was found in the 2500–2700 cm^{-1} region. Thomas [29], Ferraro *et al.* [56, 61], and Braunholtz *et al.* [84] have pointed out the presence of three broad bands at about 2500–2700 cm^{-1}, 2100–2350 cm^{-1}, and 1600–1700 cm^{-1}. These are all broad bands and of considerable width.

TABLE XII
Position of the Phosphoryl Absorption *vs* Electronegativity of Organo Groups in Several Acid Organophosphates

Phosphoric acid $(GO)_2 PO(OH)$ G Group	Electronegativity of G Group by Bell's method [33]	$\nu_{P=O}$, cm^{-1}
phenyl	8.7	1200
n-amyl		1210
isoamyl	↑	1212
isobutyl		1213
β-phenylethyl		1215
n-butyl	8.3	1220
nonyl		1225
2-ethylhexyl		1225
γ-phenylpropyl		1230
isopropyl	↓	1235
butoxyethyl		1240
cyclohexyl	7.5	1240

TABLE XIII
P=O Stretching Frequencies in Various Acidic Organophosphorus Compounds

Compound	$\nu_{P=O}$, cm^{-1}	Reference
$(\varphi O)_2 PO(OH)$	1250	16
$(\varphi CH_2 O)_2 PO(OH)$	1212	16
(p-tolyl—O)_2 PO(OH)	1195	61
(o-tolyl—O)_2 PO(OH)	1230	61
(β-phenylethyl—O)_2 PO(OH)	1215	61
(γ-phenylpropyl—O)_2 PO(OH)	1230	61
(butyl—O) phenyl—PO(OH)	1215	56
(2-ethylhexyl—O) phenyl—PO(OH)	1215	56
(butyl—O) butyl—PO(OH)	1238	56

There appears to be some controversy regarding whether all three of these bands are associated with the hydroxyl stretching and bending fundamental modes of vibration. It is generally agreed that the absorption at higher frequency is the fundamental stretching vibration [16, 29, 56, 80]. Hadži [73] believes the band at 2100 –

TABLE XIV
POH Frequencies in Several Acidic
Organic Phosphorus Acids

Compound	ν_{POH}, cm^{-1}	Reference
$(\varphi O)_2 PO(OH)$	2597, 2000–2275, 1667	16
$(\varphi CH_2 O)_2 PO(OH)$	2564, 2273, 2128, 1695	16
$(BuO)_2 PO(OH)$	2600, 2280, 1665	85
$(PrO)_2 PO(OH)$	2668, 1712	85
(butyl—O) phenyl—PO(OH)	2610, 2300, 1670	56
(2-ethylhexyl—O) phenyl—PO(OH)	2650, 2300, 1690	56
(butyl—O) butyl—PO(OH)	2600, 2300, 1690	56
(tridecyl—O) phenyl—PO(OH)	2650, 2350, 1690	56
(p-tolyl—O)_2 PO(OH)	2650, 2475, 1900, 1700	61
(o-tolyl—O)_2 PO(OH)	2700, 2350, 1700	61
(β-phenylethyl—O)_2 PO(OH)	2700, 2300, 2150, 1700	61
(γ-phenylpropyl—O)_2 PO(OH)	2600, 2300, 1700	61

2350 cm^{-1} is also associated with the fundamental stretching vibration and that the splitting at 2500–2700 cm^{-1} and 2100–2350 cm^{-1} is due to the two minima in potential energy separated by a small potential barrier, permitting tunneling of the protons. The unsymmetrical hydrogen-bonded structure of an organophosphoric acid is shown below:

$$\text{GO} \diagdown \quad \text{O} \cdots \text{HO} \quad \diagup \text{OG}$$
$$\text{P} \qquad \qquad \text{P}$$
$$\text{GO} \diagup \quad \text{OH} \cdots \text{O} \quad \diagdown \text{OG}$$

Thomas [72] believes that the 2100–2350 cm^{-1} and 1700 cm^{-1} peaks are associated with combination tones. The 2100–2350 cm^{-1} band is a combination of the δ_{OH} vibration at 1250 cm^{-1} and $\nu_{(PO)-H}$ vibration at 910–1030 cm^{-1}. The low-frequency band is a combination of the $\nu_{P=O}$ band at about 1250 cm^{-1} and the absorption at 400–500 cm^{-1}. The two higher-frequency bands shift to lower frequency on deuteration. There is some question as to the effect of deuteration on the lower-frequency band, since the 2100–2350 cm^{-1} band is shifted into this vicinity. Thomas [72] finds the δ_{OH} in POH groups to occur near 1250 cm^{-1}. The absorptions from 1600–2700 cm^{-1}

TABLE XV
Infrared Correlations for Organophosphorus Compounds

	Band, cm^{-1}	Intensity
P=O stretch – neutral compounds	1150–1350	s
P=O stretch – acidic compounds	1150–1245	s
P—H stretch	2300–2450	m
P—O—C (alkyl) stretch	950–1050	s
(P—O)—C (aryl) stretch	900–1000	s
P—(O—C) (aryl) stretch	1160–1240	m
P—C (aryl) stretch	1420–1450	m-s
P—F stretch	740–930	s
P—Cl stretch	430–585	s
P—O—P stretch	950–1000	s
P—(O—H) stretch – acidic compounds	2500–2700	w-m (broad)
(P—O—H) deformation – acidic compounds	1250	m
(P—O)—H stretch	900–1000	s

are very characteristic and can be used to identify these compounds even if the precise causes for them are unknown.

PSH Correlation

Organophosphorus compounds containing this group have only been superficially studied. McIvor studied compounds containing the PSH group from $1-2.6$ μ, and found absorptions at 5076–5247 cm^{-1}. Menefee et al., [85] reported absorption in compounds of the type $(RO)_2 PS (SH)$ at about 2440 cm^{-1}. Allen and Colclough [86] found the absorption at 2420 cm^{-1} in compounds of the type $(R)_2 PS (SH)$. The absorption appears to be at a low frequency for an SH stretch, indicating possible hydrogen bonding, and reports of studies with solvents have indicated intermolecular hydrogen bonding [85, 86]. Further work is necessary to evaluate this correlation. To date, no reports of a bending P—SH vibration have appeared.

CONCLUSIONS

The infrared correlations for organophosphorus compounds are tabulated in Table XV. These are the ranges that can be considered, at the moment, to be fairly well established for certain groups in organophosphorus compounds. As more compounds are examined new correlations will appear and the old correlations will be adjusted slightly. Three recent publications should be mentioned: the works of Thomas and Chittenden [87, 88] and the paper of Detoni and Hadži [89].

REFERENCES

1. Popov, E. M., Kabachnik, M. I., and Mayants, L. S., *Russian Chem. Rev.* 30:362 (1961).
2. *Chem. Eng. News* 30:4513 (1952).
3. Freedman, L. D., and Doak, G. O., *Chem. Rev.* 57:479 (1957); *Chem. Rev.* 61:31 (1961).
4. Frank, A. W., *Chem. Rev.* 61:389 (1961).
5. Berlin, K. D., and Butler, G. B., *Chem. Rev.* 60:243 (1960).
6. Arbuzov, A. E., Batuev, M. I., and Vinogradova, U. S., *Compt. rend.* 14:599 (1946).
7. Meyrick, C. I., and Thompson. H. W., *J. Chem. Soc.* p. 225, (1950).
8. Gore, R. C., *Discussions Faraday Soc.* 9:138 (1950).
9. Daasch, L. W., and Smith, D. C., *Anal. Chem.* 23:853 (1951); U.S.N.R. Report. No. 3657 (1950).
10. Daasch, L. W., and Smith, D. C., *J. Chem. Phys.* 19:22 (1951).
11. Holmstedt, B., and Larsson, L., *Acta Chem. Scand.* 5:1179 (1951).
12. Bergmann, E. D., Littauer, U. Z., and Pinchas, S., *J. Chem. Soc.* p. 847, (1952).
13. Bennett, F. W., Emeleus, H. J., and Haszeldine, R. N., *J. Chem. Soc.* p. 3598, (1954).
14. Harvey, R. B., and Mayhood, J. E., *Can. J. Chem.* 33:1554 (1955).
15. Bellamy, L. J., and Beecher, L., *J. Chem. Soc.* p. 476, (1952).
16. Bellamy, L. J., and Beecher, L., *J. Chem. Soc.* p. 1701, (1952).
17. Bellamy, L. J., and Beecher, L., *J. Chem. Soc.* p. 729, (1953).
18. Maarsen, J. W., Smit, M. C., and Matze, J., *Rec. Trav. Chim.* 76:713 (1957).
19. Gerding, H., and Maarsen, J. W., *Rec. Trav. Chim.* 76:481 (1957).
20. Daasch, L. W., *J. Am. Chem. Soc.* 80:5301 (1958).
21. Houalla, D., and Wolf, R., *Bull. soc. franç.* 1:129 (1960).
22. Houalla, D., and Wolf, R., *Compt. rend.* 247:482 (1958).
23. Burger, L. L., *J. Phys. Chem.* 62:590 (1958).
24. Burger, L. L., HW–44888, (1957).
25. Ketelaar, J. A. A., and Gersmann, H. R., *Rec. Trav. Chim.* 78:190 (1959).
26. Lagowski, J. J., *Quart. Rev.* 13:233 (1959).
27. Griffiths, J. E., and Burg, A. B., *J. Am. Chem. Soc.* 82:1507 (1960).
28. Schmutzler, R., *J. Inorg. & Nuclear Chem.* 25:335 (1963).
29. Thomas, L. C., *Chem. & Ind.* p. 197, (1957).
30. Thomas, L. C., and Chittenden, R. A., *Chem. & Ind.* p. 1913, (1961).
30a. Hooge, F. N., and Christen, P. J., *Rec. Trav. Chim.* 77:911 (1958).
31. Moedritzer, K., and Irani, R. R., *J. Inorg. & Nuclear Chem.* 22:297 (1961).

32. Richard, J. J., Burke, K. E., O'Laughlin, J. W., and Banks, C. V., *J. Am. Chem. Soc.* 83:1722 (1961).
33. Bell, J. V., Heisler, J., Tannenbaum, H., and Goldenson, J., *J. Am. Chem. Soc.* 76:5185 (1954).
34. Mortimer, F. S., *Spectrochim. Acta* 9:270 (1957).
35. Gordy, W., and Stanford, S. C., *J. Chem. Phys.* 8:170 (1940); 9:204 (1941).
36. Marvel, C. S., Copley, M. J., and Ginsburg, E. J., *J. Am. Chem. Soc.* 62:3109 (1940).
37. Audrieth, L. F., and Steinman, R. J., *Ibid.* 63:2115 (1941).
38. Kosolapoff, G. M., and McCullough, J. F., *Ibid.* 73:5392 (1951).
39. Halpern, E., Bouck, J., Finegold, M., and Goldenson, J., *Ibid.* 77:4472 (1955).
40. Aknes, G., and Gramstad, T., *Acta Chem. Scand.* 14:1485 (1960).
41. Gramstad, T., *Ibid.* 15:1337 (1961).
42. Bellamy, L. J., and Williams, R. L., *Trans. Faraday Soc.* 55:14 (1959).
43. Bellamy, L. J., Conduit, C. P., Pace, R. J., and Williams, R. L., *Ibid.* 55:1677 (1959).
44. Bellamy, L. J., and Rogasch, P. E., *Spectrochim. Acta* 16:30 (1960).
45. Bellamy, L. J., and Williams, R. L., *Proc. Roy. Soc. (London)*, Ser. A 255:22 (1960).
46. Ferraro, J. R., *Appl. Spectroscopy* 17:12 (1963).
47. McIvor, R. A., Grant, G. A., and Hubley, C. E., *Can. J. Chem.* 34:1611 (1956).
48. McIvor, R. A., and Hubley, C. E., *Can. J. Chem.* 37:869 (1959).
49. Popov, E. M., *J. Gen. Chem. USSR* 29:1967 (1959).
50. Herzberg, G., Infrared and Raman Spectra of Polyatomic Molecules, D. Van Nostrand, Princeton, (1945).
51. Jost, H., and Anderson, B., *J. Chem. Phys.* 2:624 (1934).
52. Daasch, L. W., *J. Am. Chem. Soc.* 80:5301 (1958).
53. Corbridge, D. E. C., and Lowe, E. J., *J. Chem. Soc.* p. 4555, (1954).
54. Beachell, M. C., and Katlafsky, B., *J. Chem. Phys.* 27:182 (1957).
55. Miller, C. D., Miller, R. C., and Rogers, W. R., *J. Am. Chem. Soc.* 80:1562 (1958).
56. Peppard, D. F., Ferraro, J. R., and Mason, G. W., *J. Inorg. & Nuclear Chem.* 12:60 (1959).
57. Nyquist, R. A., *Appl. Spectr.* Vol. 161 (1957).
58. Chapman, A. C., and Harper, R., *Chem. & Ind.* p. 985, (1962).
59. Ferraro, J. R., *J. Inorg. & Nuclear Chem.* 24:475 (1962).
60. Ferraro, J. R., Peppard, D. F., and Mason, G. W., *Spectrochim. Acta* 19:811 (1963).
61. Peppard, D. F., Ferraro, J. R., and Mason, G. W., *J. Inorg. & Nuclear Chem.* 16:246 (1961).
62. Halmann, M., *Spectrochim. Acta* 16:407 (1960).
63. Deacon, G. B., and Jones, R. A., *Australian J. Chem.* 15:555 (1962).
64. Kaesz, H. D., and Stone, F. G. A., *Spectrochim. Acta* 15:360 (1959).
65. Heisler, J., Bouck, J. B., Finegold, H., and Goldenson, J., Chemical Corps and Radiological Laboratories 310 (1954).
66. Baudler, M., *Z. Elektrochem.* 59:173 (1955).
67. Rice, R. G., Daasch, L. W., Holden, J. R., and Kohn, E. J., *J. Inorg. & Nuclear Chem.* 5:190 (1958).

68. Delwaulle, M. K., and Francois, F., *Compt. rend.* 220:817 (1945); 222:139 (1946); 224:1422 (1947).
69. Daasch, L. W., *J. Am. Chem. Soc.* 76:3403 (1954).
70. Nyquist, R. A., *Spectrochim. Acta* 19:713 (1963).
71. Moedritzer, K., and Irani, R. R., *J. Inorg. & Nuclear Chem.* 22:297 (1961).
72. Thomas, L. C., Chittenden, R. A., and Hartley, H. E. R., *Nature* 192:1283 (1961).
73. Hadži, D., and Novak, A., *Proc. Chem. Soc.* p. 241, (1960).
74. Winand, L., and Dréze, P., *Bull. Soc. Chem.* 71:410 (1962).
75. Ferraro, J. R., Mason, G. W., and Peppard, D. F., *J. Inorg. & Nuclear Chem.* 22:285 (1961).
76. Kosolapoff, G. M., and Powell, J. S., *J. Chem. Soc.* p. 3535 (1950).
77. Ferraro, J. R., and Peppard, D. F., *J. Phys. Chem.* (to be published).
78. Peppard, D. F., Ferraro, J. R., and Mason, G. W., *J. Inorg. & Nuclear Chem.* 4:371 (1957).
79. Peppard, D. F., Ferraro, J. R., and Mason, G. W., *J. Inorg. & Nuclear Chem.* 7:231 (1938).
80. Ferraro, J. R., unpublished data.
81. Ferraro, J. R., and Peppard, D. F., *Nucl. Sci. Eng.* 16:389 (1963).
82. Baes, C. F., Jr., Zingaro, R. A., and Coleman, C. F., *J. Phys. Chem.* 62:129 (1958).
83. Peppard, D. F., and Ferraro, J. R., *J. Inorg. & Nuclear Chem.* 10:275 (1959).
84. Braunholtz, J. T., Hall, G. E., Mann, F. G., and Sheppard, N., *J. Chem. Soc.* p. 868, (1959).
85. Menefee, A., Alford, D., and Scott, C. B., *J. Chem. Soc.* 25:370 (1956).
86. Allen, G., and Colclough, R. O., *J. Chem. Soc.* p. 3912, (1957).
87. Thomas, L. C., and Chittenden, R. A., *Spectrochim. Acta* 20:467 (1964).
88. Thomas, L. C., and Chittenden, R. A., *Spectrochim. Acta* 20:489 (1964).
89. Detoni, S., and Hadži, D., *Spectrochim. Acta* 20:949 (1964).

Organometallic Spectra of Silicon, Germanium, Tin, and Lead

Francis J. Bajer

Hooker Chemical Corporation
Grand Island, New York

Due to the availability and commercial importance of silicones, the infrared spectra and structural correlations of a host of organosilicon compounds have been rather extensively covered in a number of publications. Not only the conventional rock-salt region (4000–650 cm^{-1}) [1, 8] but also the far-infrared regions (650–200 cm^{-1}) [17, 18] have been covered.

Structural correlations and the infrared spectra of the organoderivatives of the remaining group IV elements, germanium, tin, and lead, on the whole have been sparse and widely scattered in the literature. The trend to correlate the spectral properties along a group in the Periodic Table has of late been increasing with the works of Noltes [10, 13–15], Rochow [36, 37, 50, 55], and Chumaevskii [23, 38, 54], to mention a few. It is the purpose of this short review to compile, although not exhaustively, group frequency correlations useful for the elucidation of certain structural features characteristic of the organometallic compounds of the group IV elements.

The interpretation of an infrared spectrum of any material is most generally based on the concept and use of the group frequency. It is well known that particular aggregates of atoms (e.g., functional groups) will absorb infrared radiation within certain frequency intervals and thus will characterize the materials as to the type (s) of groups present, thereby enabling pertinent structural features to be elucidated. Although these group frequencies are generally located within rather narrow limits, interferences or perturbations of these characteristic frequency bands can and do arise through such effects as the relative electronegativity of neighboring groups, spatial geometry of the molecule, and mechanical mixing of vibrations.

Partially because of the size and mass of the silicon atom, the infrared spectrum of an organosilicon compound has been found to be, to a first approximation, a sum of the characteristic absorptions of what are considered to be essentially independent group vibrations [1, 17]. Indeed, it is this vibrational insulation offered by the silicon atom that has enabled the organosilicon materials to be used as fairly good models for extending group frequency correlations to the other members of the group. Russian workers [56] have likewise referred to these "barrier" properties of the heavier elements, and thus a similarity of certain physical and chemical properties is understandable and to a certain degree predictable.

METAL—H ABSORPTION

Absorption data concerning the Si—H bond have been extensively studied [1, 17, 18, 23, 25, 38, 50–53].

The Si—H stretching mode generally falls within the frequency interval of 2100 to 2300 cm^{-1}. As the number of substituents bonded to the silicon atom increases, there is a corresponding shift in frequency of the Si—H stretching vibration to lower wave number. Alkyl substituents tend to shift the ν_{Si-H} to lower frequency, while aryl substituents tend to raise the frequency.

An associated Si—H bending vibration is generally found in the 750–950 cm^{-1} region. The number of substituents linked to the silicon is determinable from the position and number of intense absorptions in this region. Hence, a monosubstituted silane general-

TABLE I
ϕ_3 M—H

M	νM—H, cm^{-1}	δM—H, cm^{-1}
Si	2135	805
Ge	2040	720
Sn	1820	565

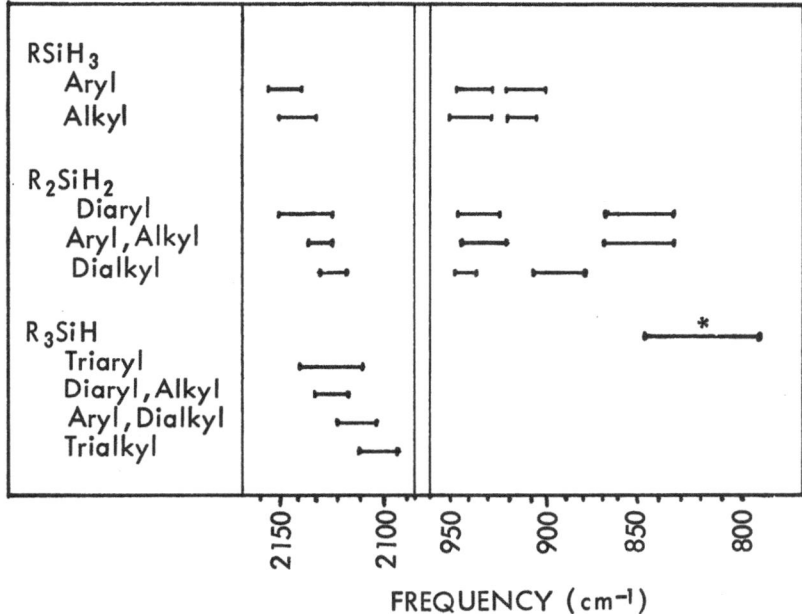

RSiH$_3$
  Aryl
  Alkyl

R$_2$SiH$_2$
  Diaryl
  Aryl, Alkyl
  Dialkyl

R$_3$SiH
  Triaryl
  Diaryl, Alkyl
  Aryl, Dialkyl
  Trialkyl

2150   2100   950   900   850   800

FREQUENCY (cm^{-1})

Fig. 1. Generalized correlation chart for the silicon–hydrogen vibrations in substituted silanes. [*The spectral interferences in this region prevented an accurate determination of the frequency range for the various types of trisubstituted silanes.] [Reprinted by permission from *Spectrochim. Acta* 15:651 (1959).]

ly shows two strong bands absorbing between 950 and 900 cm^{-1}. A disubstituted silane shows one band in the 950–900 cm^{-1} region and another strong band at about 890–825 cm^{-1}. For trisubstituted compounds, the region of absorption below 850 cm^{-1} is frequently seriously overlapped by other group absorptions and consequently is not as useful for correlation work.

The absorptions for a number of mono-, di-, and trisubstituted organosilanes are summarized in Fig. 1 [52].

The shift of the frequency of the M—H stretching and the deformation modes can be demonstrated by an examination of the spectra of a series of triphenyl–metal derivatives (Table I).

METAL—PHENYL ABSORPTION

In general, all of the phenyl—bearing elements of group IV (i.e., Si, Ge, Sn, and Pb) show striking similarity with regard to the absorption pattern due to the phenyl group. In analogy with the

TABLE II

Frequency interval.cm⁻¹	Probable assignment
3100~3000 (w-m)	aromatic C—H stretch
2000~1650 (vw)	overtone or combination bands due to C—H out-of-plane vibration
1600~1430 (m-s)	skeletal ring vibrations
1130~1000 (m-s)	C—H in-plane deformations
800~670 (m-s)	C—H out-of-plane deformations

w = weak, m = medium, s = strong, vw = very weak.

conventional organic compounds, Table II summarizes the general frequency intervals where absorption bands are found, and the most probable vibrational assignments.

The skeletal ring vibrations are displaced to lower frequencies by approximately 20 cm⁻¹ as compared to organic spectra, and there appears to be little change, if any, from these frequency positions as the group IV elements are descended, with generally three absorption bands appearing at 1580, 1480, and 1430 cm⁻¹.

Intensity differences of these bands relative to each other have been observed [1]. The band at 1430 cm⁻¹ deserves special attention because of its constancy of position and its sharpness. It has been observed for all compounds bearing an unsubstituted phenyl ring linked directly to a metal. Indeed, it has been associated with monosubstituted phenyl generally linked to silicon [2], although it also occurs with other metal—phenyl bands [5-7]. It is interesting to note that this band is absent or considerably broadened and decreased in intensity for materials where disubstituted phenyl groups are present [3, 4].

Further information regarding the presence of monosubstituted phenyl groups linked to metal atoms can be obtained from an examination of the substitution pattern region (2000~1650 cm⁻¹) and the C—H out-of-plane deformation vibration (800–700 cm⁻¹) in analogy to organic spectra [8].

The substitution pattern for monosubstituted aryl groups linked to silicon, germanium, tin, or lead is that normally characteristic for such a grouping, namely, four weak bands spread between 2000 and 1600 cm⁻¹ [9]. Associated with these bands are two strong absorptions occurring at 730 and 675 cm⁻¹ due to the C—H out-of-plane modes.

The substitution band pattern between 2000 and 1650 cm⁻¹ is not as generally applicable for the identification of position isomers as one might assume. This becomes evident when one investigates the infrared spectra of several disubstituted aryl silanes [3] (Figs. 2, 3, 4, and 5). This isomer identification (i.e., *ortho, meta,* or *para*) for a series of trimethylsilyl derivatives having the structural characteristics of (I) was apparently feasible, while structures of type (II) were not readily identifiable.

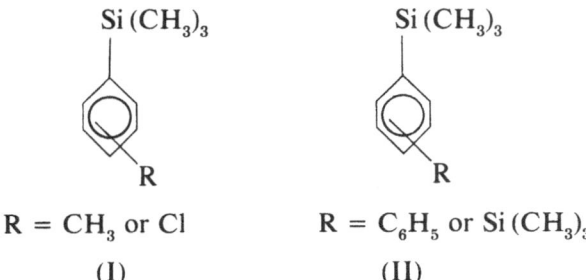

$$\text{R} = \text{CH}_3 \text{ or Cl} \qquad\qquad \text{R} = \text{C}_6\text{H}_5 \text{ or Si}(\text{CH}_3)_3$$

$$(\text{I}) \qquad\qquad\qquad (\text{II})$$

Differentiation among the phenyl-bearing congeners of group IV is possible in the 1125–1050 cm⁻¹ region. This correlation, first noted by J. G. Noltes *et al.* [10], was observed in a series of organometallic tetraphenyls. For the series Si to Pb, the following strong absorption bands appeared to be characteristic of the metal–phenyl band; 1100 cm⁻¹ (Si—ϕ), 1080 cm⁻¹ (Ge—ϕ), 1065 cm⁻¹ (Sn—ϕ), and 1052 cm⁻¹ (Pb—ϕ). This vibration has been ascribed to a perturbed phenyl vibration and is seemingly analogous to the band shift occurring in a family of aromatic halides as the mass of the halogen varies from that of F to that of I [11].

The strong intensity and relative constancy characteristic of each metal–phenyl band has been of value in elucidating the structures of a series of mixed organometallic compounds of group IV [12–15]. In compounds where two different elements are linked to phenyl groups, the spectra show bands distinctive for each (Fig. 6).

In a more intensive study of some twenty phenyltin compounds by Poller [16], the 1065 cm⁻¹ band characteristic for Sn—ϕ has been found to vary in its frequency position depending on the substituents linked to tin, and consequently overlapping of the Ge—ϕ position at 1080 cm⁻¹ or the Pb—ϕ frequency at 1052 cm⁻¹ is possible.

In addition to the work done within the rock-salt region of the spectrum (4000–650 cm⁻¹), far-infrared correlations are being expanded [17–20].

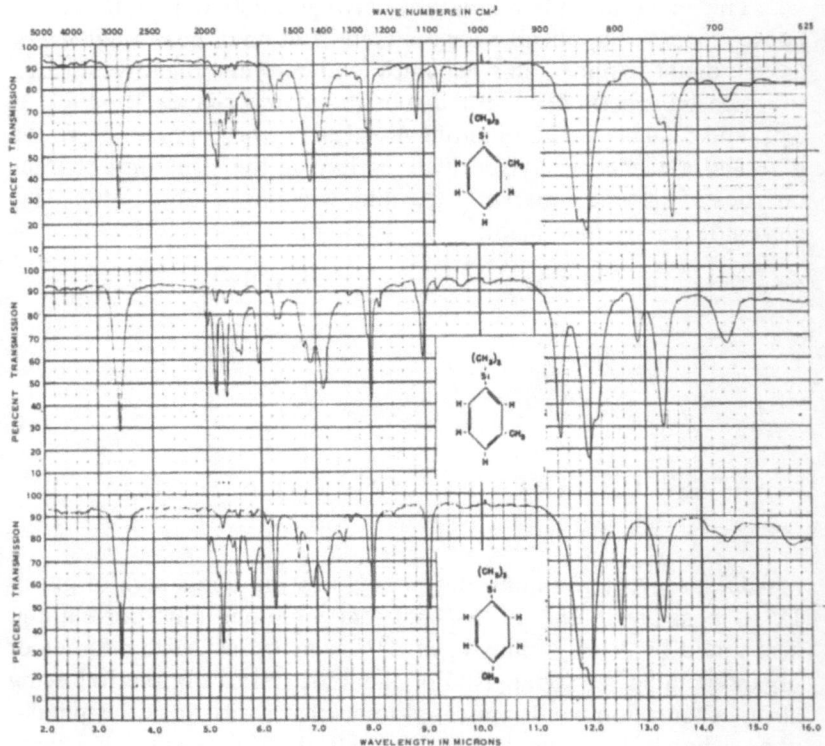

Fig. 2. Infrared spectra of *o*, *m*, *p*-methyl-substituted trimethylsilylbenzenes. [Reprinted by permission from *J. Am. Chem. Soc.* 73:3798 (1951).]

METAL–VINYL AND METAL–ALLYL ABSORPTIONS

The presence of a metal atom in a position adjacent to a site of unsaturation has a pronounced effect on both the intensity and frequency of absorption. This can readily be seen on investigating a series of vinyl or allyl compounds.

In a series of vinyl compounds studied by Henry and Noltes [15], Table III reveals a gradual shift of the C—H out-of-plane vibrational mode of the vinyl group as the series is descended within the 960–940 cm⁻¹ region.

In a comparable series of allyl compounds, where the—HC= CH₂ group is insulated through a methylene unit, two absorption bands appear at 930 cm⁻¹ and 894 cm⁻¹, but no shift of these bands has been noticed in descending the series.

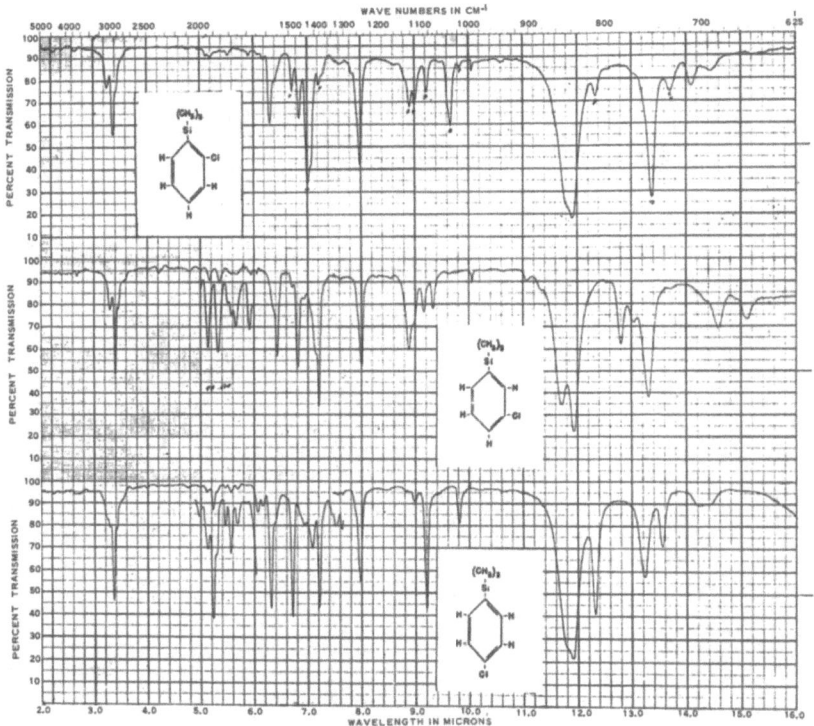

Fig. 3. Infrared spectra of *o, m, p*-chloro-substituted trimethylsilylbenzenes. [Reprinted by permission from *J. Am. Chem. Soc.* 73:3798 (1951).]

Intensity and frequency changes involving the C=C stretching region between 1650 and 1600 cm^{-1} has been investigated by Mironov and Chumaevskii [54].

For a series of organosilicon vinyl compounds of the type $Cl_x(CH_3)_{3-x} SiCH=CH_2$, the $\nu_{C=C}$ occurs at about 1600 cm^{-1}.

TABLE III

Compound	C—H out-of-plane, cm^{-1}
$\phi_3 SiCH=CH_2$	960
$\phi_2 Si(CH=CH_2)_2$	960
$\phi_3 Ge\,CH=CH_2$	952
$\phi_2 Ge(CH=CH_2)_2$	952
$\phi_3 Sn\,CH=CH_2$	950
$\phi_2 Sn(CH=CH_2)_2$	950
$\phi_3 Pb\,CH_2=CH_2$	942

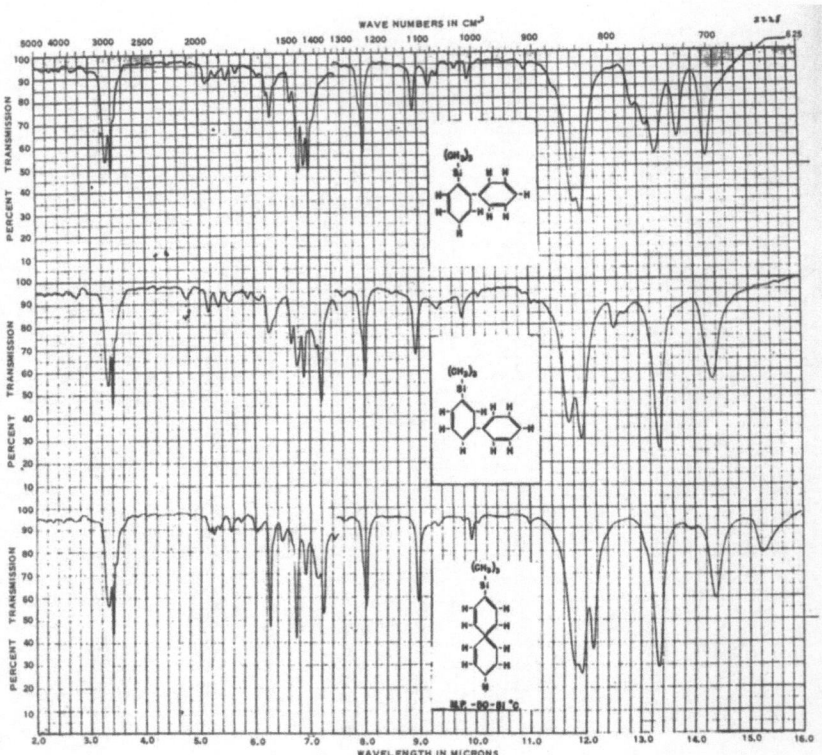

Fig. 4. Infrared spectra of *o, m, p*-phenyl-substituted trimethylsilylbenzenes. [Reprinted by permission from *J. Am. Chem. Soc.* 73:3798 (1951).]

However, the intensity of this absorption increases (by a factor of three) as x increases. For the corresponding allyl series, i.e., $Cl_x(CH_3)_{3-x} SiCH_2 CH{=}CH_2$, the $\nu_{C=C}$ shifts to higher frequency by about 35 cm^{-1}, but the intensity relationship is reversed from that reported for the vinylic compounds.

Table IV shows the effect on the position of the C═C stretching frequency for the series $(CH_3)_3 MCH_2 CH{=}CH_2$ as the group is descended. Along with the shift of the $\nu_{C=C}$ frequency there is a gradual increase in the intensity of absorption.

The insulating effect of the methylene group interposed between the silicon atom and the C═C bond can be seen for the series $(CH_3)_3 Si (CH_2)_x CH{=}CH_2$ (Table V), with the $\nu_{C=C}$ frequency approaching that usually found in isolated carbon–carbon double bonds.

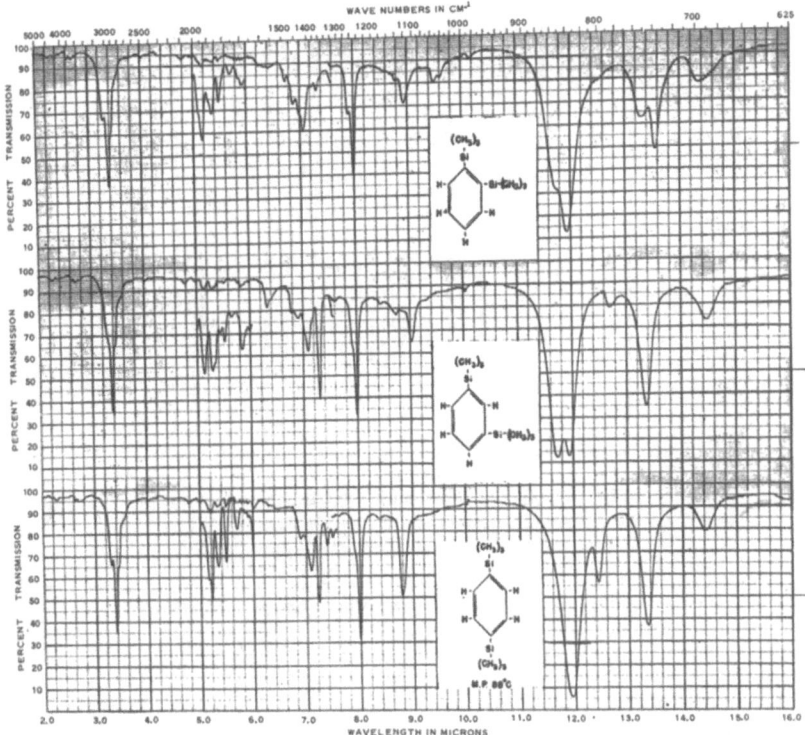

Fig. 5. Infrared spectra of o, m, p-trimethylsilyl-substituted trimethylsilylbenzenes. [Reprinted by permission from *J. Am. Chem. Soc.* 73:3798 (1951).]

The shift of the carbon–carbon double bond frequency as well as the change in intensity of the absorption band is most probably associated with the change in dipole moment made possible for such elements as silicon, germanium, tin, etc., as a result of $d_\pi - p_\pi$ bond-

TABLE IV

$(CH_3)_3\ MCH_2\ CH{=}CH_2$	
M	$\nu_{C=C}, cm^{-1}$
C	1650
Si	1633
Ge	1633
Sn	1628

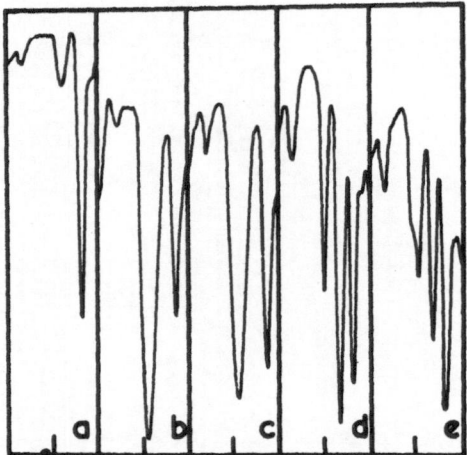

Fig. 6. Infrared absorption spectra (8.5–9.5 μ) of (a) $Ph_3SnCH_2CH_2SnPh_3$; (b) $Ph_3SnCH_2SiPh_3$; (c) $Ph_3SnCH_2CH_2Si(Ph_2)CH_2CH_2SnPh_3$; (d) $Ph_3SnCH_2CH_2GePh_3$; and (e) $Ph_3SnCH_2CH_2Ge(Ph_2)CH_2CH_2SnPh_3$. [Reprinted by permission from *J. Am. Chem. Soc.* 82:558 (1960).]

ing. Such interaction of the p–electron orbitals of unsaturated systems with the vacant d–orbitals of the group IV elements has been used to explain frequency shifts to some extent, not only for material containing olefinic groups, but also for shifts occurring in the polyphenyl derivatives [57], α–silyl and germyl ketones [58], metallo–isocyanate compounds [59] and for a series of ethynyl–silanes and stannanes [60, 61] (Table VI).

TABLE V

$(CH_3)_3 Si (CH_2)_x CH{=}CH_2$	
x	$\nu_{C=C}$, cm^{-1}
0	1598
1	1633
2	1645
3	1648

TABLE VI
Frequency Shifts Partially Attributable to $p_\pi - d_\pi$ Bonding

Compound	$\nu_{C=O}$	$\nu_{C=C}$	$\nu_{C\equiv C}$	Reference
ϕ_3 CCO ϕ	1692	— —	— —	59
ϕ_3 Si CO ϕ	1618	— —	— —	59
ϕ_3 Ge CO ϕ	1629	— —	— —	59
Me_3 CC≡C CH=CH_2	— —	1611	2204	62
Me_3 Si C≡C CH=CH_2	— —	1612, 1599	2147	62
Me_3 Sn C≡C CH=CH_2	— —	1605	2129	62
Me_3 CC≡CH	— —	— —	2105, 2135	61
Me_3 Si C≡CH	— —	— —	2035	61
Me_3 Sn C≡CH	— —	— —	2010	61
ϕ_3 Si C≡CH	— —	— —	2037	61

METAL – OXYGEN ABSORPTION

The silicon – oxygen bond has received wide attention and has been studied by many workers both in the Raman and infrared fields [1, 2, 8, 17, 21–33]. Although this list is far from complete, it does take into account many interesting structural features which can be deduced from infrared spectral examination.

The asymmetric Si—O stretching frequency normally lies between 1100 and 1000 cm⁻¹. Wright [21] and Hunter [2] were among the first to show the utility of the siloxane (Si—O—Si) absorption bands in this region for the characterization of linear and cyclic polymers.

In view of the large percentage of ionic character inherent in the silicon – oxygen bond, absorption associated with this linkage proves to be considerably more intense than, for example, the conventional ether link (C—O—C), and is generally five times as intense and, therefore, rather easily identified.

Although the infrared spectra show a certain amount of congruence with increasing molecular weight for both the cyclic and linear polysiloxanes, nevertheless the spectral pattern for the Si—O bond is significantly different to permit identification of the siloxane as either straight-chain or cyclic (Figs. 7 and 8).

For cyclic siloxanes, the trimer generally absorbs at 1010–1020 cm⁻¹, while the tetramer absorbs in the 1070–1090 cm⁻¹

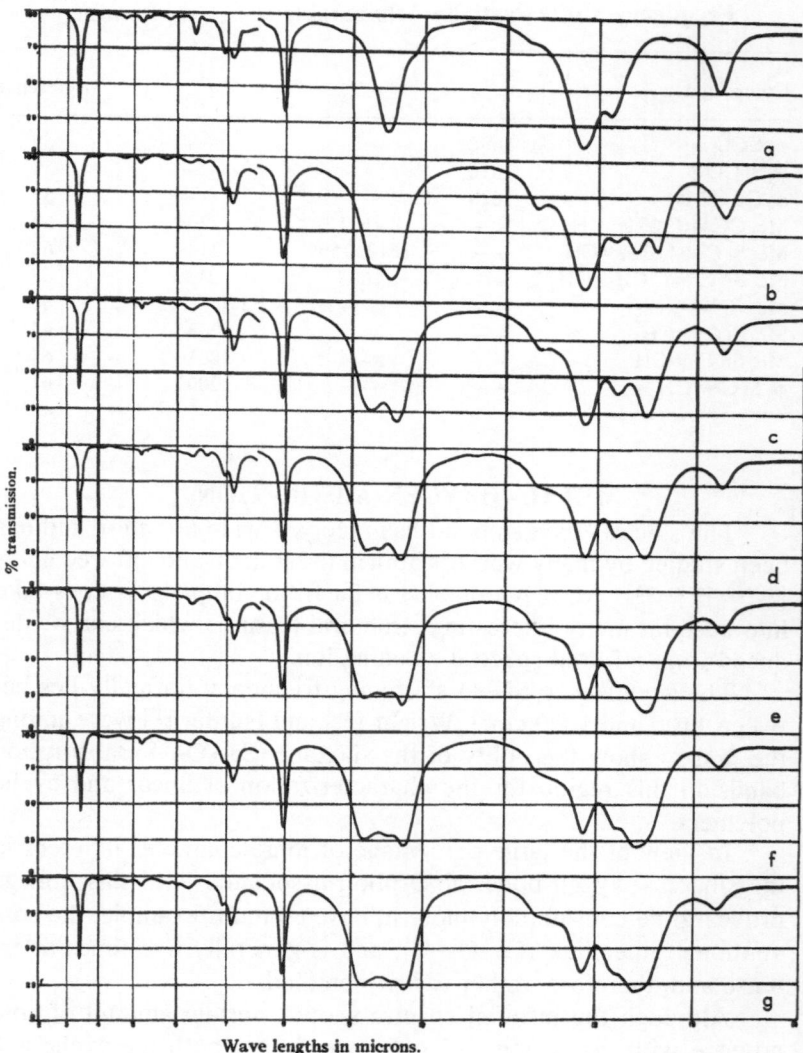

Wave lengths in microns.

Fig. 7. Infrared spectra of linear polymethylsiloxanes. (a) Hexamethyldisiloxane,
(b) octamethyltrisiloxane, (c) decamethyltetrasiloxane, (d) dodecamethylpenta-
siloxane, (e) tetradecamethylhexasiloxane, (f) hexadecamethylheptasiloxane,
(g) octadecamethyloctasiloxane. [Reprinted by permission from *J. Am. Chem. Soc.*
69:803 (1947).]

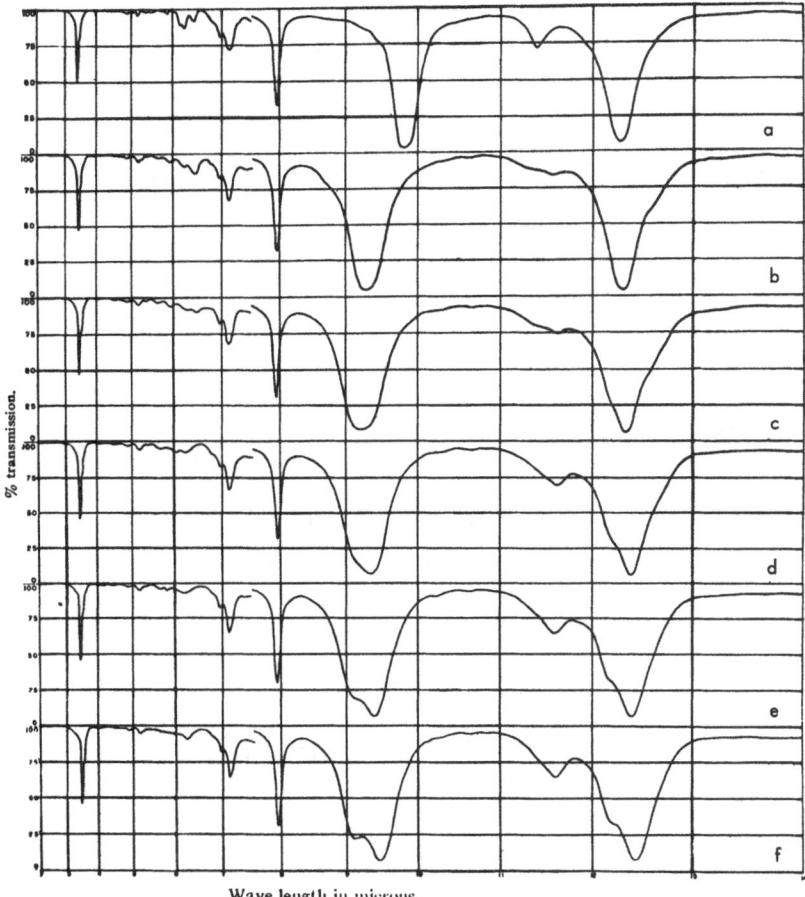

Wave length in microns.

Fig. 8. Infrared spectra of cyclinc dimethylsiloxanes. (a) Hexamethylcyclotrisil-oxane, (b) octamethylcyclotetrasiloxane, (c) decamethylcyclopentasiloxane, (d) dodecamethylcyclohexasiloxane, (e) tetradecamethylcycloheptasiloxane, (f) hexa-decamethylcyclo-octasiloxane. [Reprinted by permission from *J. Am. Chem. Soc.* 69:803 (1947).]

region. The shift in frequency of approximately 65 cm^{-1} in going from the planar trimer to the nonplanar, eight-membered tetramer has been attributed to ring strain [1]. Differentiation among the higher-membered rings (i.e., pentamer, hexamer, etc.) becomes increasingly more difficult for although there is a shift toward lower frequency, the differences are small, and moreover the absorption pattern overlaps that of the linear polymers.

TABLE VII

Compound	Asymmetric, cm⁻¹	Symmetric, cm⁻¹
$(CH_3)_3$ Si—O—Si $(CH_3)_3$	1050	520
$(Cl)_3$ Si—O—Si $(Cl)_3$	1110	– – –

The linear polysiloxanes tend to absorb at about 1050 cm⁻¹ for the lower-molecular-weight materials, showing a gradual spread between 1100 and 1000 cm⁻¹ and eventually exhibiting a broad, strong absorption with maxima located at approximately 1080 cm⁻¹ and 1030 cm⁻¹.

An insight into the explanation for the behavior of the siloxane shift can be gained by examining the work of Noll [32]. It is generally known that the frequency position of any particular band is influenced by either a change in the force constant of the band, or a change in the masses in proximity to the bond under investigation, or a change in the particular bond angle. Any and all of these parameters will cause an absorption band to shift in either of two directions, to higher or lower frequencies. Thus, the introduction of electron-withdrawing groups or electron-donating groups can and does cause the Si—O bond absorption to shift to higher or lower frequencies, depending of course on whether these effects tend to strengthen or weaken the bond strength. Such an effect can be exemplified by carefully analyzing both the asymmetric and symmetric Si—O stretching frequencies (Table VII).

It is interesting to note, however, that differences in electronegativities or differences in masses of the attached groups seem to shift both the asymmetric and symmetric Si—O band in the same direction (either to higher or to lower frequency) but not to the same degree; a change in bond angle causes a divergence in the two bands. Table VIII summarizes the bond angles for a representative number of siloxanes.

Other factors being equal, it seems reasonable to assume that since the valence bond angle of the cyclic trisiloxanes are generally smaller than those of the cyclic tetrasiloxanes, the trimer will generally absorb at lower frequencies than the tetramer as regards the asymmetric Si—O stretching frequency (the reverse is true for the

TABLE VIII

Compound	<Si—O—Si	Reference
[(CH$_3$)$_2$SiO]$_4$	142.5	34, 32
[(CH$_3$)$_2$SiO]$_3$	136	34
	125	32
[(C$_2$H$_5$)$_3$SiO]$_2$	130 ± 5	29
[(CH$_3$)$_3$SiO]$_2$	130 ± 10	32
	137 ± 7	32
[Cl$_3$SiO]$_2$	175 ± 5	32

symmetric mode). In all cyclic trimers and tetramers, this effect is clearly borne out [1, 2, 21, 26, 35].

In a series of cyclic tetrasiloxanes containing a functional group linked to silicon, the ν_{as} Si—O stretching frequency remained relatively constant through the 1080–1089 cm^{-1} region [27].

For a series of spiro-type silicon trimers and tetramers, the characteristic shift of the siloxane absorption for trimer and tetramer was similarly observed (Fig. 9) [35].

Other ring systems involving the Si—O bond reveal further frequency shifts as examples of ring-strain effects (Table IX).

The Ge—O bonds have not been studied as exhaustively as has the Si—O bond. However, papers have appeared which tentatively assign the asymmetric Ge—O stretching vibration to a region ranging from 700 to 900 cm^{-1} [36–39].

The absorption in this region of the spectrum is marred by the fact that Ge—CH$_3$ rocking vibrations overlap those of the Ge—O link [37]. It is also interesting to note that for the cyclic trimer and tetramer, although distinguishable, the shift is considerably smaller (i.e., approximately 12 cm^{-1} as compared to about 65 cm^{-1} for Si—O), with the Ge—O—Ge bond absorbing at 848 cm^{-1} for the trimer, 860 cm^{-1} for the tetramer, and 868 cm^{-1} for the higher-molecular-weight polymer [36, 37] (see Fig. 10).

As with the Ge—O bond, references to Sn—O bonds are sparsely scattered throughout the literature [37, 40]. In the case of the heavy tin atom, vibrations involving the stretching mode of the Sn—O bond lie in the region from 580 to 650 cm^{-1} [37], although in many cases bands at about 780 cm^{-1} [41] have been observed.

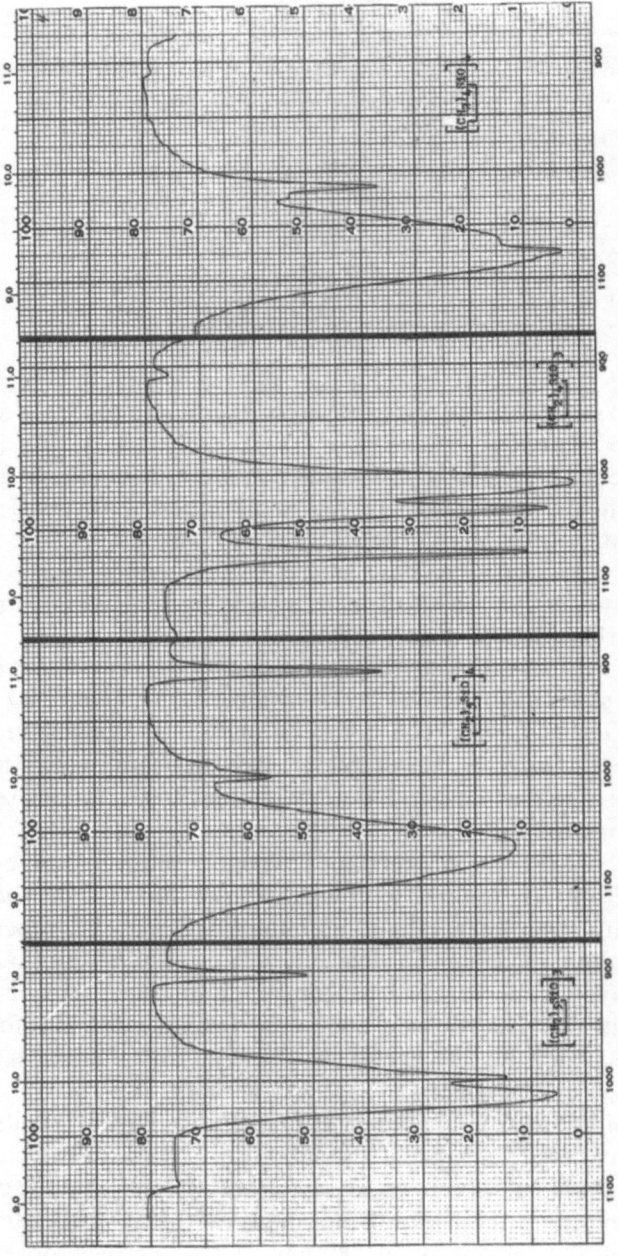

Fig. 9. CS$_2$ solution spectra of silacyclohexane and silacyclopentane, tri- and tetracylclosiloxanes.

TABLE IX

Ring system	ν_{as}Si—O, cm^{-1}	Reference
![ring system 1] $R_2Si\diamond SiR_2$ with CH_2 CH_2 and Si R_2	1000–1040	33
$\phi_2Si\diamond Si\phi_2$ with ϕ_2Si—$Si\phi_2$	955	28
$R_2Si\diamond SiR_2$ with CH_2—CH_2	925	30

In any case, the Sn—O absorption seems to border on the limits of the rock-salt region (650 cm^{-1}) and is best observed as an intense, broad absorption in the KBr or CsBr regions (Figs. 11 and 12).

Again, very little has apparently been accomplished in studying the spectra of organolead oxides. Where data are available [40] it is assumed that the Pb—O link absorbs at about 625 cm^{-1}. Such appears to be the case for $[(C_2H_5)_3Pb]_2O$, where this band is assigned to be the asymmetric stretching vibration of the Pb—O bond. Not enough information is available, however, to determine a useful set of correlations.

METAL–NITROGEN ABSORPTIONS

Silicon–nitrogen-containing compounds and the characteristic absorptions involving specifically the N—H stretching and deformation modes when this unit is attached to a silicon atom, have been reported in the literature [1, 43-47, 49].

In the case of primary silicon amines (e.g., —Si NH$_2$), two

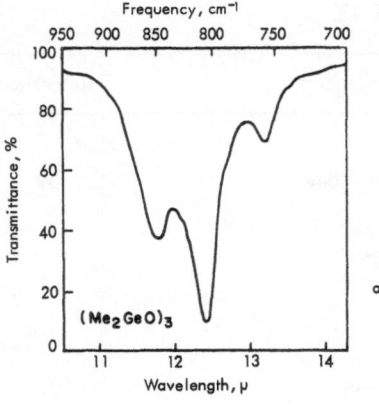

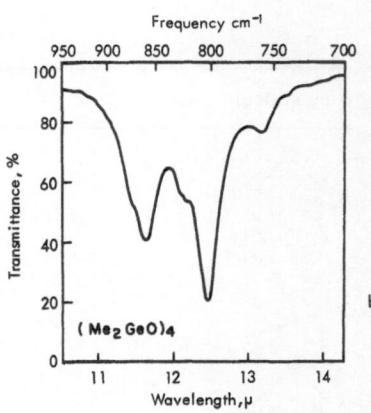

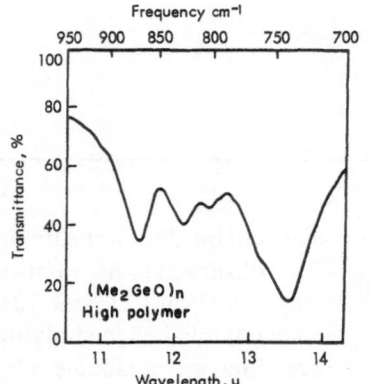

Fig. 10. Infrared spectra of di-methylgermanium oxides. [Reprinted by permission from *J. Am. Chem. Soc.* 82:4166 (1960).]

sharp but relatively weak absorption bands are found in the 3570–3380 cm⁻¹ region, while for a secondary type of silicon amine, only one weak band at about 3390 cm⁻¹ is evident [1, 44, 45]. These bands are due to the asymmetric and symmetric N—H stretching frequencies.

The bending frequencies have been found to run at lower wave numbers, with bands appearing at 1530 cm⁻¹ for primary silyl amines and at 1180 cm⁻¹ for secondary amines. These bands are fairly strong.

A strong absorption band occurring at or near the 900 cm⁻¹ region appears to be due to the asymmetric stretching mode of the Si—N—Si system [1, 44, 45].

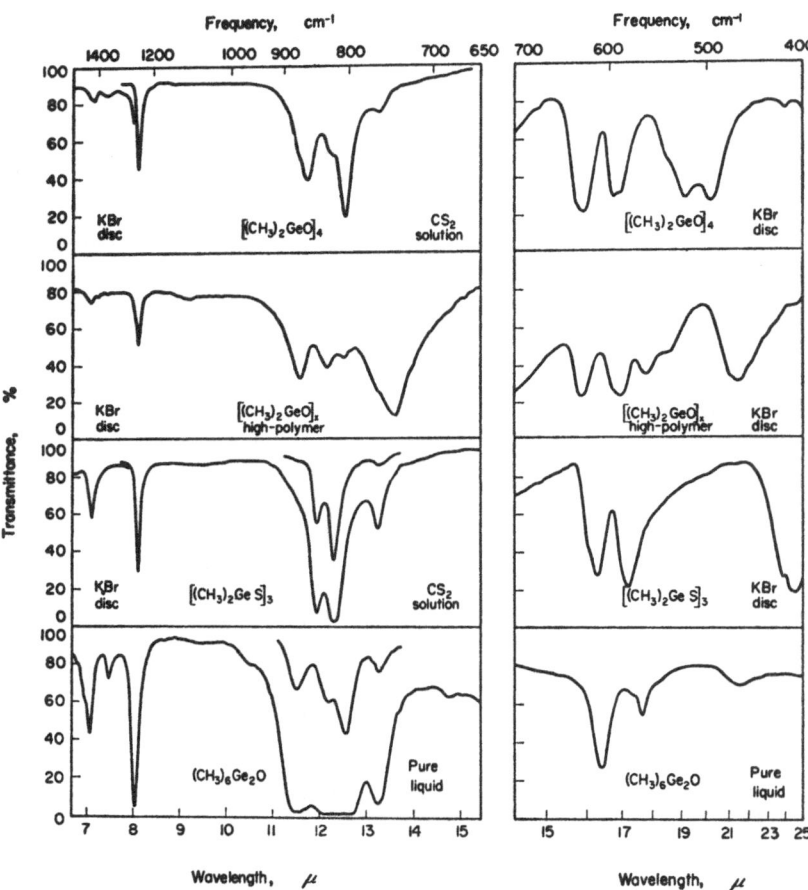

Fig. 11. Infrared spectra of methylgermanium oxides. [Reprinted by permission from *Spectrochim. Acta* 16:595 (1960).]

As in the case of the cyclic siloxanes previously discussed where ring compounds are involved (i.e., cyclic silazanes), there appears to be a characteristic shift of the asymmetric Si—N band which is dependent on the size of the silazane ring. Hence, cyclic trisilazanes show a ν_{as} Si—N—Si at 920 cm^{-1}, while the corresponding cyclic tetrasilazanes absorb at 940 cm^{-1}. A second band, attributable to the N—H deformation, shows a similar shift in

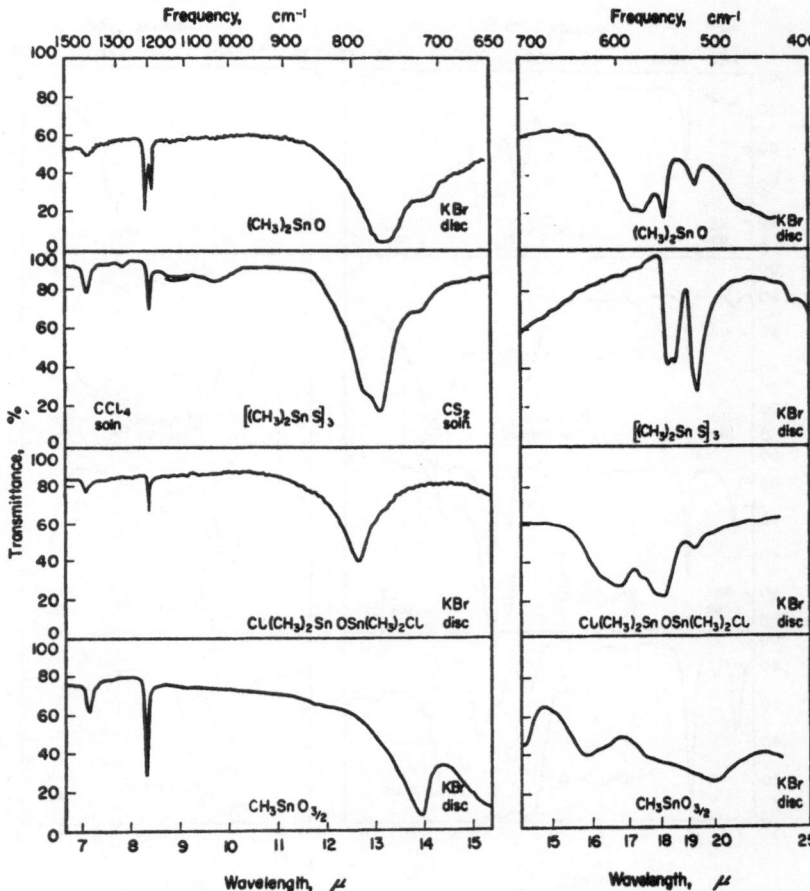

Fig. 12. Infrared spectra of methyltin oxides. [Reprinted by permission from *Spectrochim. Acta* 16:595 (1960).]

frequency, absorbing at 1150 cm⁻¹ for the trimer and about 1180 cm⁻¹ for the tetramer [1]. The magnitude of the shift, some 20 cm⁻¹, is not as pronounced as that for a similar set of cyclic siloxanes, which show a frequency shift of some 60 cm⁻¹ in going from trimer to tetramer.

Again, the study of highly strained silazane ring systems has shown that the abnormal frequency shifts of the Si—N band can be explained in much the same manner as the shift in siloxanes. For example, the ν_{as} Si—N of a cyclic disilazane absorbs at a frequency

TABLE X

	ν_{as} Si—N, cm^{-1}	δN—H, cm^{-1}	Reference
H | N / \\ R$_2$Si SiR$_2$ | | R$_2$Si SiR$_2$ \\ / N | H	925	1141	47
R$_3$Si SiR$_2$ \\ N—H / R$_3$Si SiR$_2$	920	1172	47

of some 70–80 cm^{-1} lower than that observed for the corresponding tetramer [43].

This shift in frequency to lower wave number as the Si—N bond angle is decreased is analogous to that reported in a series of strained siloxanes.

There seems to be an added difference between the absorption pattern of cyclic and noncyclic silyl amines with regard both to the asymmetric Si—N stretch and the N—H deformation mode. Table X reveals that although there is little difference in the Si—N stretching mode, there is a distinct difference in both the intensity and position for the N—H bending vibration.

Furthermore, in N—silyl-containing silazanes of the type

there appears to be a strong absorption in the $1005-1040$ cm^{-1} region which shows an increase in intensity with an increase in the number of N—silyl groups, and which appears to be entirely absent in both hexamethylcyclotrisilazane and octamethylcyclotetrasilazane [46].

Information concerning the characteristic absorptions involving the other group IV members in their attachment to nitrogen is sparse. The only reference cited in this regard [48] lists absorption bands which seem to be characteristic for the Sn—N (Et)$_2$ group. Five bands were common to each of three individual organotin compounds bearing this group. They occurred at 1290, 1173, 1010, 880, and 780 cm^{-1}. Exposure to air resulted in a loss of absorption at these frequencies when the diethylaminotin compound was converted to the corresponding organotin carbonate.

HETEROCYCLIC RING VIBRATIONS

Incorporation of the group IV elements into ring systems which are analogous to the cyclohexane or cyclopentane series has produce some interesting correlations. Oshesky and Bentley [62] have tentatively assigned a band at approximately 910 cm^{-1} as being particularly characteristic for a series of disubstituted silacyclohexanes of the type

Another band closely associated with the 910 cm^{-1} band appears at about 995 cm^{-1} and seems to decrease in intensity as heavier alkyl groups (R) are bonded to the heteroatom, silicon.

Work done on a series of diphenylated, heterocyclic six-membered rings incorporating germanium, tin, and lead in addition to silicon has shown that this pair of bands is also evident with a change of the heteroatom from Si to Pb [5, 12, 63]. Although the absorption band at 910 cm^{-1} appears to be invariant, the band centered at about 990 cm^{-1} shifts to lower frequency with a change in the heteroatom, the most pronounced change occurring in changing from germanium to tin.

Further reduction of ring size to the corresponding five-membered rings causes the appearance of an apparently new set of absorptions occurring at approximately 1075 cm^{-1} and 1025 cm^{-1} [63]. Additional data on the absorptions associated with silacyclo-hexane and silacyclopentane systems have been published in several Russian papers [64-66].

In addition, it is interesting to note that the bands which are tentatively assigned as being due to the presence of a heterocyclic six-membered ring system are not found to occur in cyclic systems of the type*

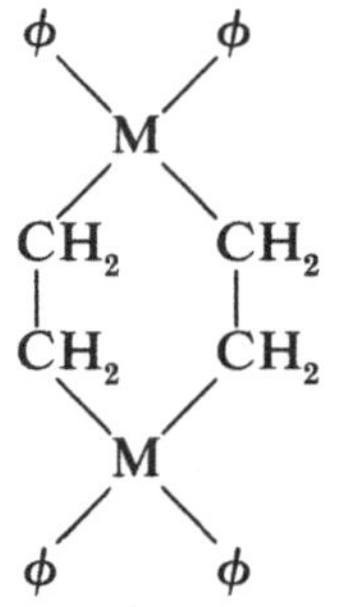

$$
\begin{array}{ccc}
\phi\diagdown\quad\diagup\phi & & CH_3\diagdown\quad\diagup CH_3 \\
M & & C \\
CH_2\quad CH_2 & or & CH_2\quad CH_2 \\
CH_2\quad CH_2 & & CH_2\quad CH_2 \\
M & & Sn \\
\phi\diagup\quad\diagdown\phi & & n\text{-}Bu\qquad n\text{-}Bu
\end{array}
$$

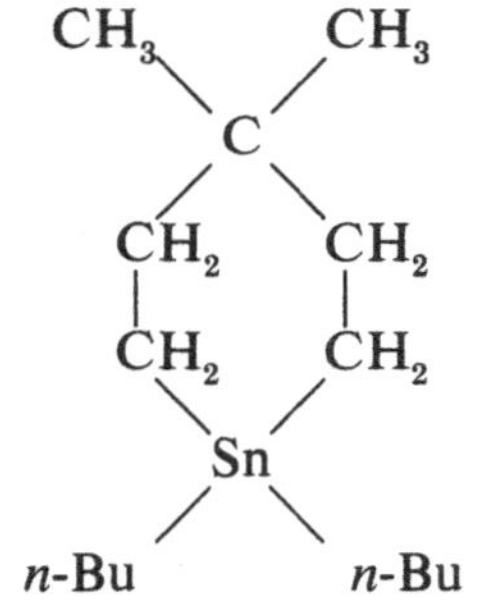

where M is Si, Ge, or Sn.

The change in size on going from the silicon atom to carbon and its effect on the carbonyl absorption have been demonstrated by Benkeser [67] by an examination of cyclohexanone analogs:

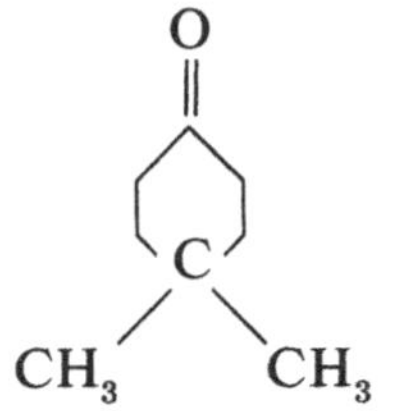

$$
\begin{array}{cc}
O & O \\
\| & \| \\
\text{(ring)} & \text{(ring)} \\
C & Si \\
CH_3\quad CH_3 & CH_3\quad CH_3 \\
\nu_{C=O}\ 1709\ cm^{-1} & \nu_{C=O}\ 1702\ cm^{-1}
\end{array}
$$

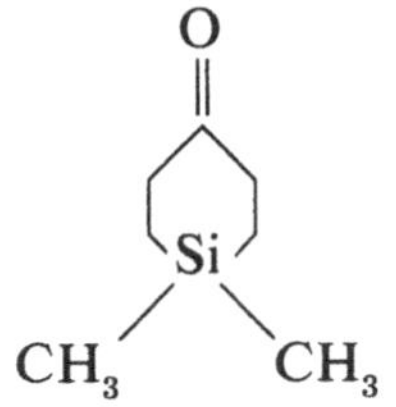

The shift in frequency to lower wave number is in the direction anticipated and generally observed by several workers in studying carbonyl frequency of alicyclic ketones as ring size is increased from five-membered to seven-membered rings.

*Spectra provided in private communication.

174 Francis J. Bajer

Further ring assignments, in particular for silicon-containing rings of varying size, have been reported in the Russian literature [65, 66].

REFERENCES

1. Smith, A. L., *Spectrochim. Acta* 16:87 (1960).
2. Young, C. W., Servais, P. C., Currie, C. C., and Hunter, M. J., *J. Am. Chem. Soc.* 70:3758 (1948).
3. Clark, H. A., Gordon, A. F., Young, C. W., and Hunter, M. J., *J. Am. Chem. Soc.* 73:3798 (1951).
4. Spialter, L., Priest, D. C., and Harris, C. W., *J. Am. Chem. Soc.* 77:6227 (1955).
5. Bajer, F. J., and Post, H. W., M. A. Thesis, The Synthesis and Infrared Study of Organometallic Heterocyclic Compounds of Group IVa, University of Buffalo, (1961).
6. Noltes, J. G., Personal communication.
7. Poller, R. C., *J. Inorg. Nucl. Chem.* 24:593 (1962).
8. Bellamy, L. J., The Infrared Spectra of Complex Molecules, second ed., New York, John Wiley & Sons, (1958).
9. Young, C. W., DuVall, R. B., and Wright, N., *Anal. Chem.* 23:709 (1951)
10. Noltes, J. G., Henry, M. C., and Janssen, M. J., *Chem. & Ind.* p. 298, (1959).
11. Kross, R. D., and Fassel, V. A., *J. Am. Chem. Soc.* 77:5858 (1955).
12. Bajer, F. J., and Post, H. W., *J. Org. Chem.* 27:1422 (1962).
13. Henry, M. C., and Noltes, J. G., *J. Am. Chem. Soc.* 82:561 (1960).
14. Henry, M. C., and Noltes, J. G., *J. Am. Chem. Soc.* 82:558 (1960).
15. Henry, M. C., and Noltes, J. G., *J. Am. Chem. Soc.* 82:555 (1960).
16. Poller, R. C., *J. Inorg. & Nuclear Chem.* 24:593 (1962).
17. Smith, A. L., *Spectrochim. Acta* 19:849 (1963).
18. Behnke, F. W., and Tamborski, C., Technical Documentary Report No. *ASD-TDR*-62-224, Directorate of Materials and Processes, Aeronautical Systems Division, Air Force Systems Command, Wright-Patterson Air Force Base, Ohio.
19. Harvey, M. C., and Nebergall, W. H., *Appl. Spectroscopy* 16:12 (1962).
20. Grenoble, M. E., and Launer, P. J., *Appl. Spectroscopy* 14:85 (1960).
21. Wright, N., and Hunter, M. J., *J. Am. Chem. Soc.* 69:803 (1947).
22. Smith, A. L., and McHard, J. A., *Anal. Chem.* 31:1174 (1959).
23. Chumaevskii, N. A., *Optics and Spectroscopy* 10(1):33 (1961).
24. Kriegsmann, H., *Z. anorg. u. allgem. Chem.* 298:223 (1959).
25. Kriegsmann, H., *Z. anorg. u. allgem. Chem.* 299:138 (1959).
26. Andrianov, K. A., Volkova, L. M., and Tartakovskaya, L. M., *Izvest. Akad. Nauk SSSR*, Otdel, Khim. Nauk 2:294 (1963).
27. Tsitsishvili, G. V., Bagratishvili, G. D., Andrianov, K. A., Khananashvili, L. M., and Kantariya, M. L., *Izvest. Akad. Nauk SSSR, Otdel. Khim. Nauk* 1014 (1962).
28. Jarvie, A. W. P., Winkler, H. J. S., and Gilman, H., *J. Org. Chem.* 27:614 (1962).
29. Kirei, G. G., and Lisitsa, M. P., *Optics and Spectroscopy* 11:28 (1961).

30. Piccoli, W. A., Haberland, G. G., and Merker, R. L., *J. Am. Chem. Soc.* 82:1883 (1960).
31. Kriegsmann, H., *Z. anorg. u. allgem. Chem.* 298:232 (1959).
32. Noll, W., *Angew. Chem.* internat. Ed. 2(2):73 (1963).
33. Andrianov, K. A., and Yakushkina, S. E., *Izvest. Akad. Nauk SSSR, Otdel. Khim. Nauk* 1396 (1962).
34. Takano, T., Kosai, N., and Kakudo, M., *Bull. Chem. Soc. Japan* 36(5):585 (1963).
35. Bajer, F. J., and Post, H. W., *J. Org. Chem.* 28:1941 (1963).
36. Brown, M. P., and Rochow, E. G., *J. Am. Chem. Soc.* 82:4166 (1960).
37. Brown, M. P., Okawara, R., and Rochow, E. G., *Spectrochim. Acta* 16:595 (1960).
38. Chumaevskii, N. A., *Optics and Spectroscopy* 13:37 (1962).
39. Metlesics, W., and Zeiss, H., *J. Am. Chem. Soc.* 82:3324 (1960).
40. Vyshinskii, N. N., and Rudnevskii, N. K., *Optics and Spectroscopy* 10:421 (1961).
41. Kochkin, D. A., and Chirgadze, Yu. N., *Zhur. Obshchei Khim.* 32:4007 (1962)
42. Fessenden, R., and Fessenden, J. S., *Chem. Rev.* 61:361 (1961).
43. Lienhard, K., and Rochow, E. G., *Angew. Chem.* internat. Ed. 2(6):325 (1963).
44. Fessenden, R., *J. Org. Chem.* 25:2191 (1960).
45. Andrianov, K. A., *et al., Izvest. Akad. Nauk SSSR, Otdel. Khim. Nauk* 1197 (1962).
46. Fink, W., *Helv. Chim. Acta* 45:1081 (1962).
47. Wannagat, U., and Brandstatter, D. O., *Angew. Chem.* internat. Ed. 2(5):263 (1963).
48. Sisido, K., and Kozima, S., *J. Org. Chem.* 27:4051 (1962).
49. Andrianov, K. A., *et al., Izvest. Akad. Nauk SSSR, Otdel, Khim. Nauk* 948 (1962).
50. West, R., and Rochow, E. G., *J. Org. Chem.* 18:303 (1953).
51. Kaye, S., and Tannenbaum, S., *J. Org. Chem.* 18:1750 (1953).
52. Kniseley, R. N., Fassel, V. A., and Conrad, E. E., *Spectrochim. Acta* 15:651 (1959).
53. Janz, G. J., and Mikawa, Y., *Bull. Chem. Soc. Japan* 34:1495 (1961).
54. Mironov, V. F., and Chumaevskii, N. A., *Doklady Akad. Nauk SSSR* 146:1117 (1962).
55. Okawara, R., Webster, D. E., and Rochow, E. G., *J. Am. Chem. Soc.* 82:3287 (1962).
56. Leites, L. A., Egorov, Yu. P., Zueva G. Ya., and Ponomarenko, V. A., *Izvest. Akad. Nauk SSSR, Otdel. Khim. Nauk* 2132 (1961).
57. Rao, C. N. R., Ramachandran, J., and Balasubramian, A., *Can. J. Chem.* 39:171 (1961).
58. Brook, A. G., Quigley, M. A., Peddle, G. J. D., Schwartz, N. V., and Warner, C. M., *J. Am. Chem. Soc.* 82:5102 (1960).
59. Miller, F. A., and Carson, G. L., *Spectrochim. Acta* 17:977 (1961).
60. West, R., and Krahanzel, C. S., *J. Inorg. Chem.* 1:967 (1962).
61. Yakovleva, T. J., Petrov, A. A., and Zavgorodnii, V. S., *Optics and Spectroscopy* 12:106 (1962).
62. Oshesky, G. D., and Bentley, F. F., *J. Am. Chem. Soc.* 79:2057 (1957).

63. Bajer, F. J., and Post, H. W., to be published.
64. Vdovin, V. M., *et al.*, *Izvest. Akad. Nauk SSSR, Otdel. Khim. Nauk* 274 (1963).
65. Egorov, Yu. P., *et al.*, *Izvest. Akad. Nauk SSSR, Otdel. Khim. Nauk* 822 (1963).
66. Vdovin, V. M., *et al.*, *Izvest. Akad. Nauk SSSR, Otdel. Khim. Nauk* 1127 (1962).
67. Benkeser, R., and Bennett, E. W., *J. Am. Chem. Soc.* 80:5414 (1958).

Arsenic Trichloride as a Solvent for Infrared Spectroscopy

Herman A. Szymanski, Robert Ripley, Robert Fiel, William Kinlin, Leon Zwolinski, Henry Drew, Dennis Bakalik, Jorge Muller, Albert Bluemle, and William Collins

Canisius College
Buffalo, New York

We shall report here the properties of arsenic trichloride as a solvent for infrared spectroscopy. We have also investigated its properties as a solvent for proton magnetic resonance spectroscopy and find it quite suitable, and where appropriate we will present the PMR spectra to verify the conclusions obtained from the infrared spectra.

We shall show that dry arsenic trichloride is an excellent solvent for covalent compounds, while arsenic trichloride containing up to 0.5% water is an excellent solvent for water-soluble compounds. We will also show that hydrogen-bonded substances are partially dissociated when dissolved in water–arsenic trichloride solutions.

GENERAL PROPERTIES OF ARSENIC TRICHLORIDE

The halide arsenic trichloride exhibits properties unlike related metal halides: it can be steam distilled; it forms no basic salt; it hydrolyzes very slowly [1]; and the hydrolysis is reversible. We have found that when water is added to arsenic trichloride two layers form and mild refluxing is required to obtain solution. The equilibrium amount of water in arsenic trichloride at 25°C is approximately 0.5% by weight. Adding more water results in a cloudy solution, indicating the presence of an insoluble oxide. While many equilibria must exist in a water–arsenic trichloride solution, we have found most of the solution behavior can be explained by assuming it is a strong acid–water solution. We shall utilize this suggestion later.

Gutman [2] has examined arsenic trichloride as a solvent for ionic reactions. Pertinent to our work was his suggestion that the halide is a highly ionic medium. This is in general agreement with our work, although our data indicate that the slight trace of water usually present in arsenic trichloride accounts for the ionic species present in the halide.

INFRARED SPECTRA OF WATER AND HYDROGEN CHLORIDE IN ARSENIC TRICHLORIDE

Since water in arsenic trichloride produces the hydrolysis product hydrogen chloride, the spectra of water and hydrogen chloride in arsenic trichloride will be discussed together. Reference will be made to the spectra of Figs. 1, 2, 3, 4, and 5.

It is possible to obtain concentrations of water up to 0.5% in arsenic trichloride without any precipitate forming. In addition, gaseous hydrogen chloride is soluble in these water–arsenic trichloride solutions but not in dry arsenic trichloride. We shall imagine that a simple reversible equilibrium reaction occurs, as follows [1]:

$$AsCl_3 + H_2O \rightleftharpoons H_3O^+ Cl^- + As(OH)_3$$

In a later section we shall show that all water–arsenic trichloride solutions are highly acidic. Heating such solutions drives off HCl gas and the equilibrium is driven to the right, with the eventual formation of an arsenic acid which in turn releases HCl. The following spectra are illustrative of this behavior. The spectra in Fig. 1 represent a water–arsenic trichloride solution of 0.5% in a cell of 0.2-mm thickness. The top spectrum is the solution as first prepared. The next spectrum is the same solution 2.5 days later; it indicates that no change has occurred on standing. The last two spectra are of the solution when it has been refluxed and the HCl has been allowed to escape.

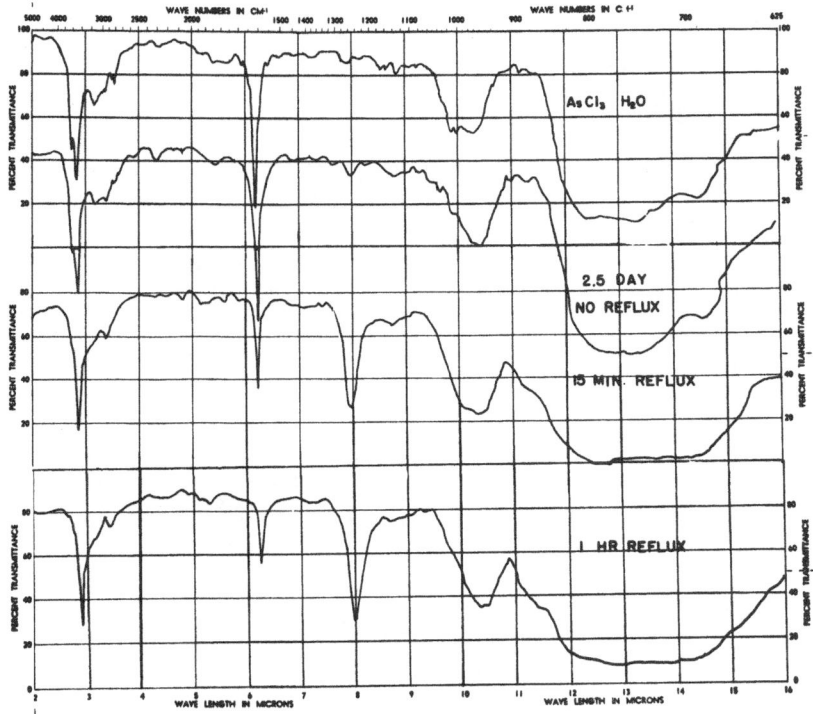

Fig. 1. $AsCl_3 - H_2O$ solutions. Top curve is freshly prepared solution; others are as marked.

In these four spectra of Fig. 1 the following assignments are made:

1. Partially dissociated water has bands at 3680, 3550, and 1600 cm⁻¹.
2. The H_3O^+ ion has bands at 3000, 1900, and 1250 cm⁻¹.
3. A band always found when water is present in arsenic trichloride is at 980 cm⁻¹.
4. Arsenic trichloride bands appear at 800 and 750 cm⁻¹. The 800 cm⁻¹ band appears on cell windows exposed to arsenic trichloride (probably an AsO band), and has medium intensity in the solution. Its intensity varies in the following spectra since the cell windows may have this band.
5. The arsenic acid has bands at 3380 and 1250 cm⁻¹.
6. Covalent HCl has a band at 2780 cm⁻¹ which is only seen when a concentrated HCl solution is examined.

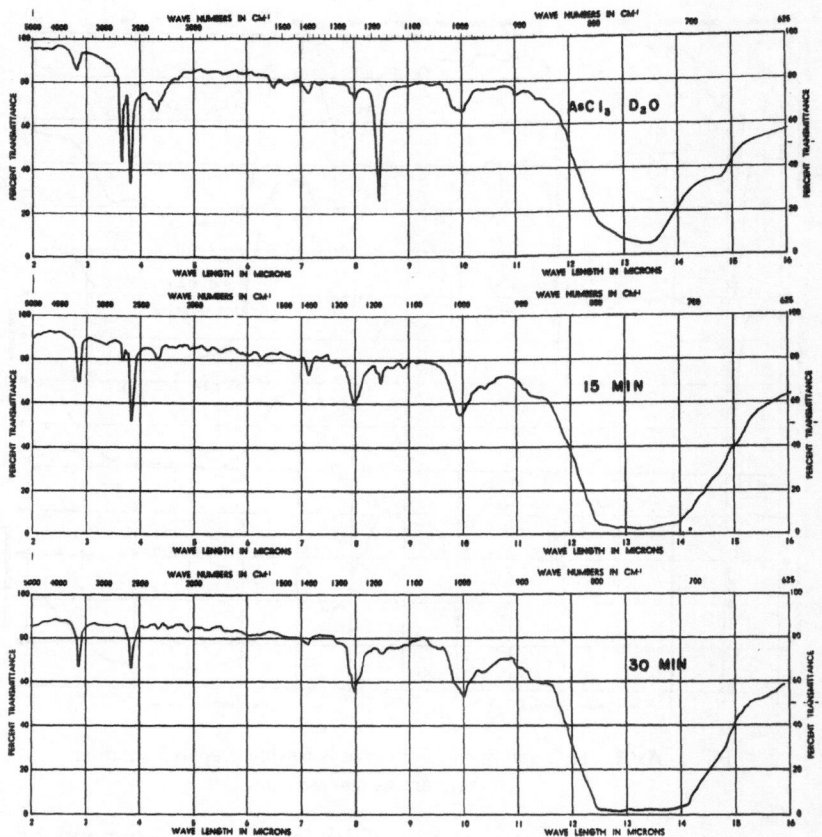

Fig. 2. AsCl$_3$–D$_2$O solutions. Top curve is freshly prepared solution;
others are as marked.

To verify these assignments the spectra in Figs. 2, 3, 4, and 5
can be examined. The spectra in Fig. 2 are of D$_2$O in arsenic trichloride, and the bands associated with H$_2$O in the spectra of Fig. 1
shift in the spectra of Fig. 2. A series of refluxes of the D$_2$O solution
was carried out, and the lower two spectra in Fig. 2 illustrate the
loss of DCl and the formation of the arsenic acid. Some H$_2$O is
picked up in the refluxing, and bands appear indicating that the acid
is not completely the deuterated compound. The cell thickness in
the spectra of Fig. 2 is the same as that in the spectra of Fig. 1.

The spectrum shown in Fig. 3 is that of a water concentration
less than 0.1%. In the spectra of Fig. 1 the water was present at
0.5%. The greater concentration of water in the spectra of Fig. 1

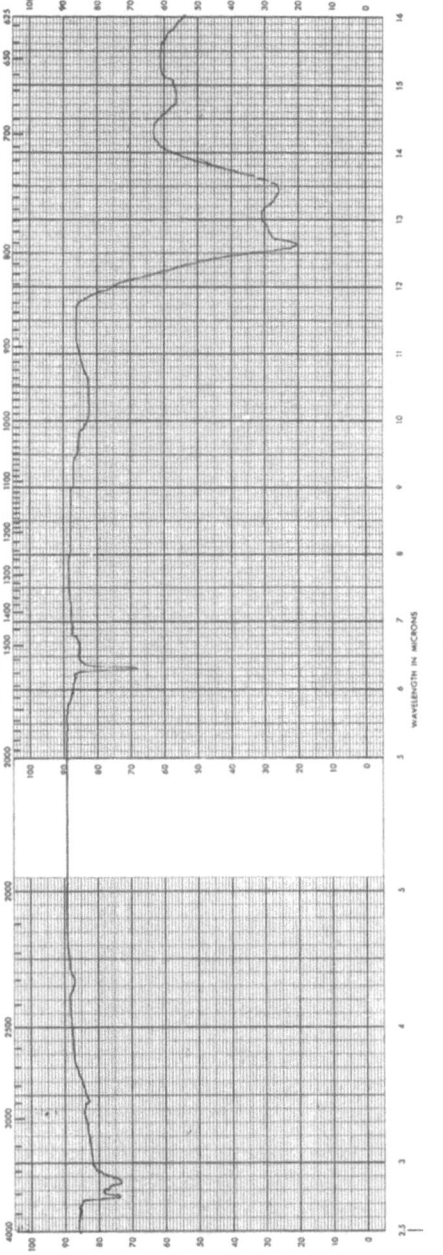

Fig. 3. $AsCl_3 - H_2O$ solution.

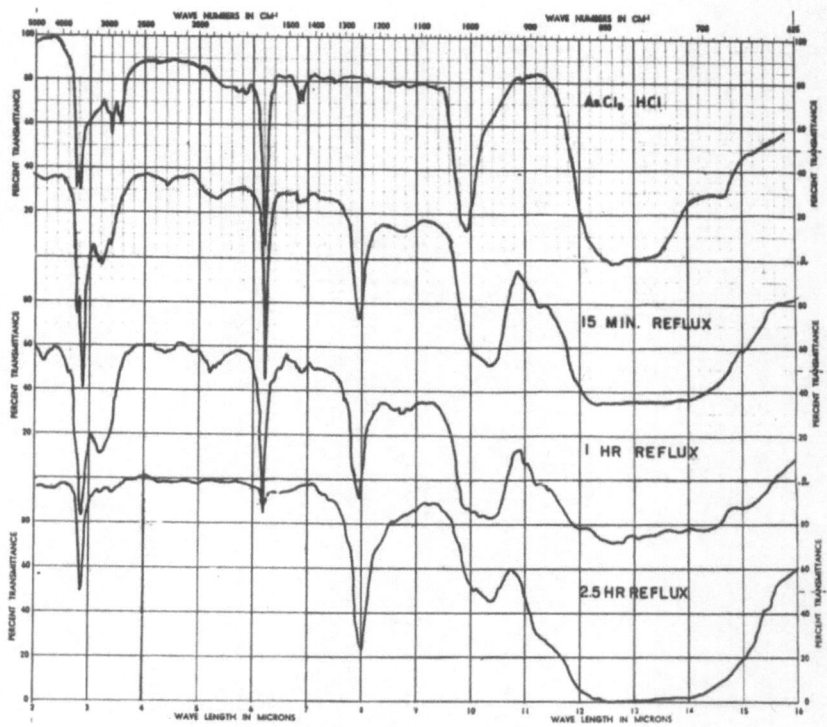

Fig. 4. AsCl$_3$–HCl solutions.

gives a strong band in the 1000 cm^{-1} range as well as a much stronger broad band in the 800–675 cm^{-1} range when Figs. 1 and 3 are compared. The 1250 cm^{-1} band suggests that some arsenic acid is also present. The greater intensity of the water bands also is indicative of the higher water concentration here. The spectrum in Fig. 3 was determined on a spectrophotometer having very good resolution in the 3000 cm^{-1} region. Two distinct bands are found for H$_2$O in this region, at 3680 and 3550 cm^{-1}. The band near 800 cm^{-1} will show variable intensity for reasons explained above.

When HCl gas is bubbled into water–arsenic trichloride solutions, it dissolves and at saturation a solution is obtained which gives spectra as shown in Fig. 4. The top spectrum is that of the saturated solution and the lower spectra are for various reflux times as the HCl gas is allowed to escape. The spectra in Fig. 5 can be compared to those in Fig. 4. Those in Fig. 5 represent a concentrated hydrochloric acid solution added to arsenic trichloride.

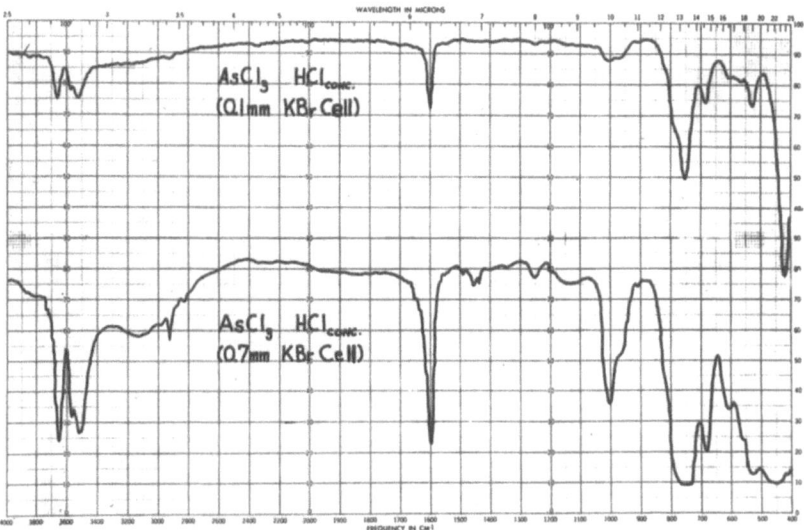

Fig. 5. AsCl$_3$–HCl solution. Cell thickness 0.1 mm (top) and 0.7 mm (bottom).

The spectra in Fig. 5 agree with the second spectrum in Fig. 4, as expected for a solution where the HCl concentration is not at saturation.

Several conclusions can be drawn from these spectra, one important one concerning the relative position of the 1000 cm^{-1} band. At the highest HCl concentration it is near 1005 cm^{-1}. As the HCl is lost it shifts to 1000 cm^{-1} with a band near 980 cm^{-1} appearing. The 980 cm^{-1} eventually becomes the only band in this region as all of the HCl is given off. This is illustrated by the last spectrum in Fig. 4 and also by the last one in Fig. 1.

The 1000 cm^{-1} band also appears when all types of solutions are freshly prepared, suggesting that until some HCl is lost the position of the band remains near 1000 cm^{-1}, an indication of high acidity. The presence of covalent HCl is indicated by the bands at 2780 and 2920 cm^{-1}, which are assigned to the P and R branches of the HCl fundamental. These bands disappear as the HCl is lost, and a broad band near 3100 cm^{-1} becomes more prominent. This band is not easily explained since other bands can appear under different circumstances in this region. For example, if we examine the spectra presented in Figs. 6, 7, 8, and 9 it can be seen that the 3100 cm^{-1} band is present in Figs. 7 and 9 and a

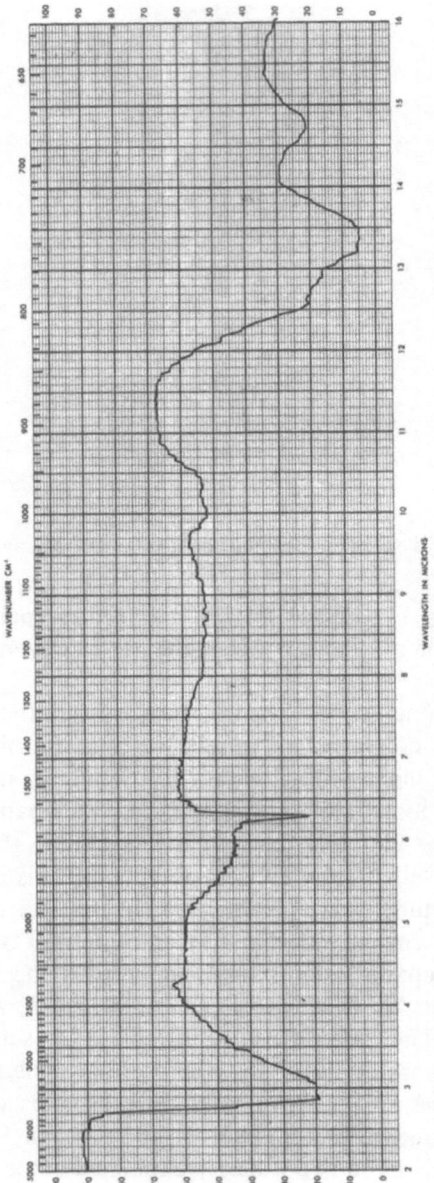

Fig. 6. AsCl$_3$–KOH solution, freshly prepared.

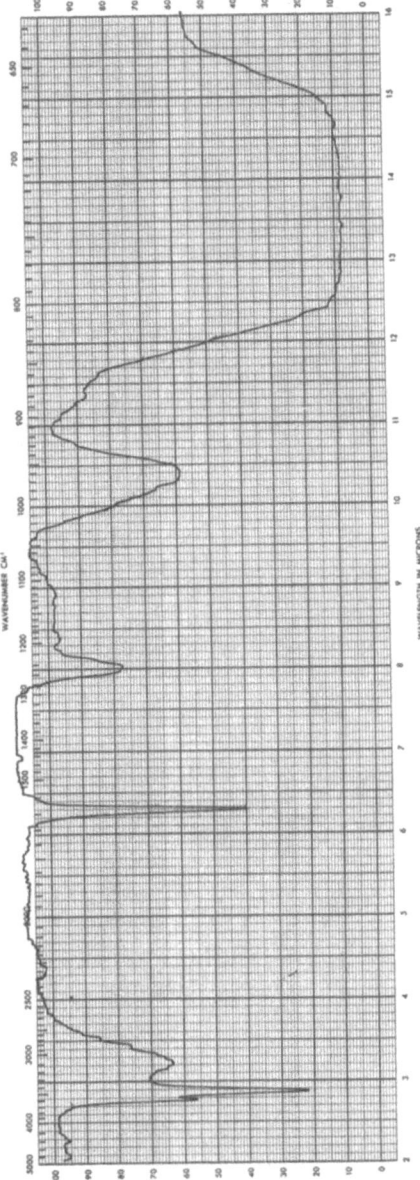

Fig. 7. AsCl$_3$ – KOH solution, 19 hr after preparation.

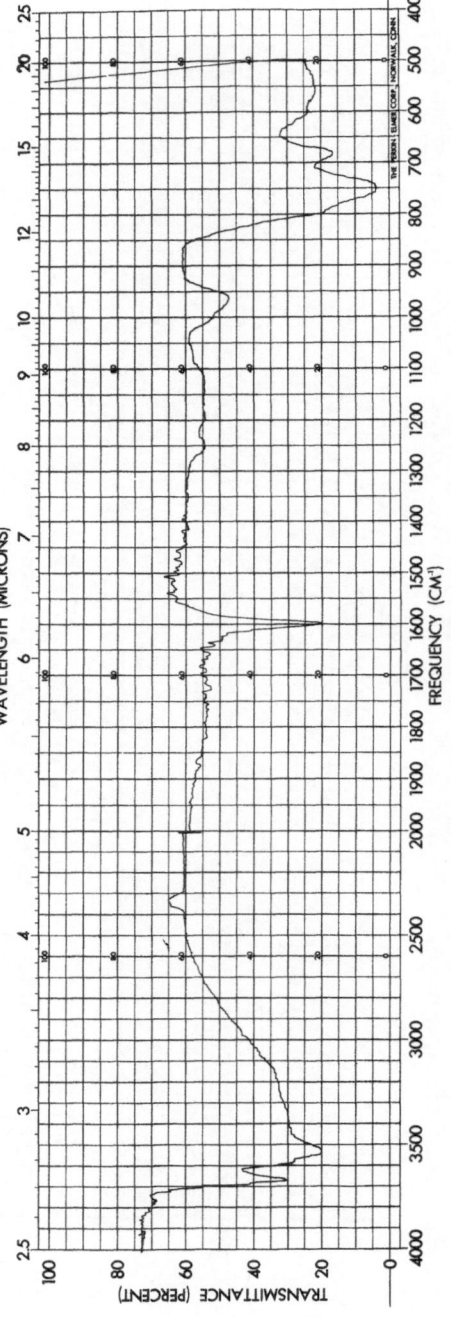

Fig. 8. AsCl₃–KOH solution, freshly prepared.

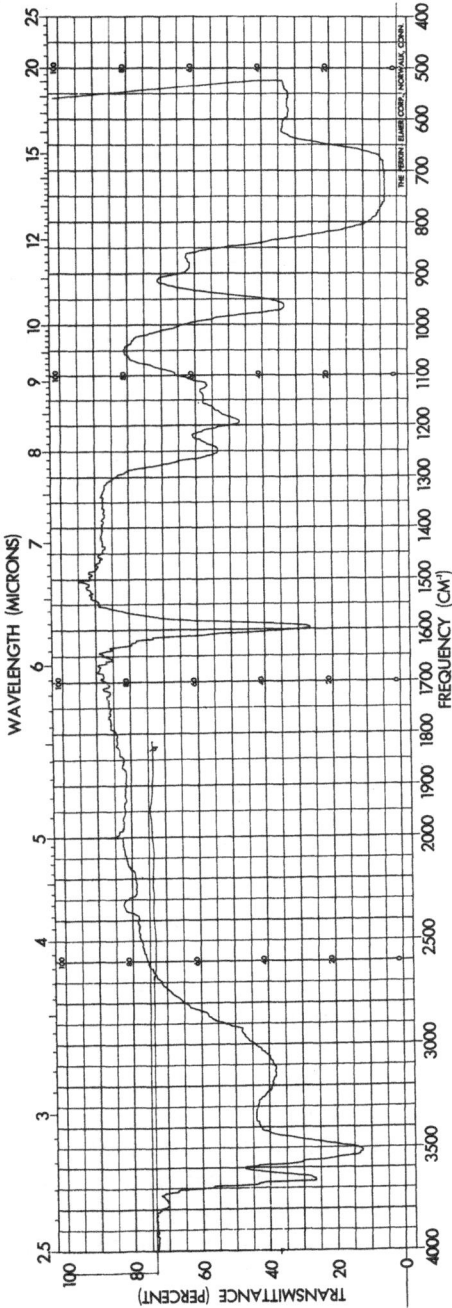

Fig. 9. AsCl$_3$–KOH solution, 48 hr after preparation.

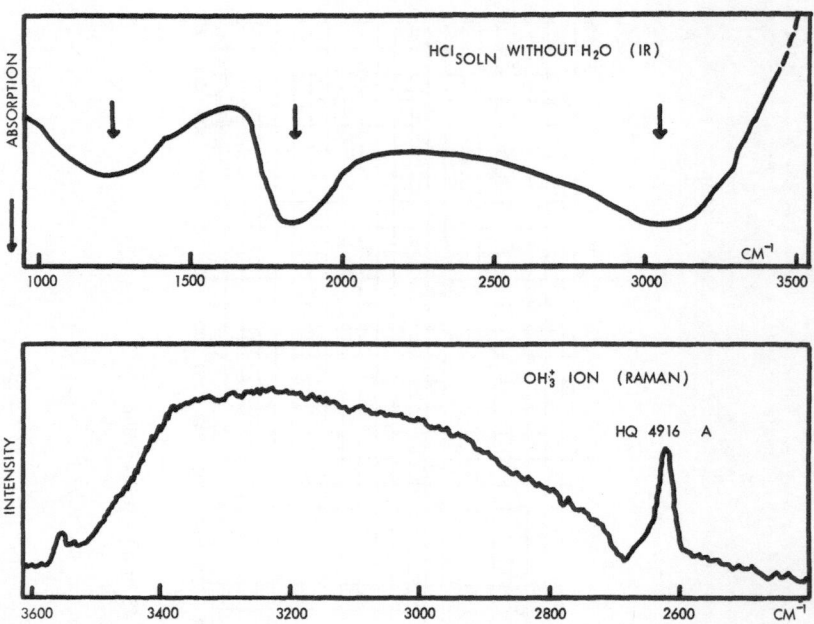

Fig. 10. Spectra of H_3O^+. Top curve is differential spectrum of a concentrated HCl solution minus H_2O. Bottom is Raman trace.

broader band centering near 3200 cm⁻¹ is present in Figs. 6 and 8. The spectra in these figures are of solutions of KOH in arsenic trichloride at a cell thickness of 0.1 mm. Figures 6 and 8 are the spectra of solutions as they are first prepared, while Figs. 7 and 9 are spectra of the same solutions after standing at least 24 hr. It is obvious that a slow reaction is occurring. We have found that water is only slowly soluble in arsenic trichloride. It is known that the halide can be steam distilled. All chemical evidence suggests that the hydrolysis is both very slow and reversible, and we conclude that the spectra in Figs. 6 and 8 represent the beginning of this hydrolysis reaction. The broad band near 3200 cm⁻¹ is merely that of liquid H_2O. As the hydrolysis continues and the concentration of HCl builds up, the water molecules are dissociated by the acid and bands due to the hydronium ion appear. It is significant that adding KOH always gives the spectra shown in Figs. 6 and 8, while adding HCl always gives spectra similar to those in Figs. 7 and 9. The 980 cm⁻¹ band shifts to 1000 cm⁻¹ in freshly prepared solutions in both instances.

The spectrum of H_3O^+ has never been clearly defined. The most clearly defined infrared and Raman spectra previously reported are shown in Fig. 10 [3, 4]. The broad and poorly defined bands are due to H_2O association and interference. In our solutions the dissociated H_2O allows the H_3O^+ bands to be more clearly defined and our assignments for this ion are given above.

We find the same spectra when HBr is added to arsenic trichloride as when HCl is added, as would be expected if our species is primarily H_3O^+.

Bands in the 1250–1100 cm^{-1} region sometimes appear as hydrolysis is continued, and these we assign to the AsO groups of hydrolysis products. The spectrum in Fig. 9 is typical for these bands.

In some instances two strong bands near 1000 and 910 cm^{-1} are found in $AsCl_3$–H_2O systems. See for example the spectrum in Fig. 11. These bands will be better explained after the spectra of carboxylic acids are discussed in a later section.

THE SPECTRA OF SULFURIC ACID
IN ARSENIC TRICHLORIDE

The behavior of sulfuric acid in arsenic trichloride is similar to that of hydrogen chloride, except that in the PMR spectrum we have not been able to observe the weak low-field signal but only the high, strong water signal. The PMR spectrum shows the effect of increasing acid concentration on the degree of association of the water better than does the infrared. The infrared spectra of various concentrations of sulfuric acid in arsenic trichloride are shown in Fig. 12, while the PMR spectra of the same solutions are shown in Fig. 13.

The infrared spectra of the various solutions show only the increase in intensity of the water vibrations as the amount of water is increased (i.e., the normality of acid is lower), and the increase in intensity of the 3100 cm^{-1} band. As previously suggested, the position of the 980 cm^{-1} band is dependent on the relative amount of HCl present from the hydrolysis and for one solution (4.45 N) the band is at 1000 cm^{-1} as it is for all freshly prepared solutions having a large amount of HCl.

The shift to higher fields of the H_2O signal in the PMR spectra as the acid concentration is increased is an indication of the increased dissociation of H_2O. No weak signal is found on the low-

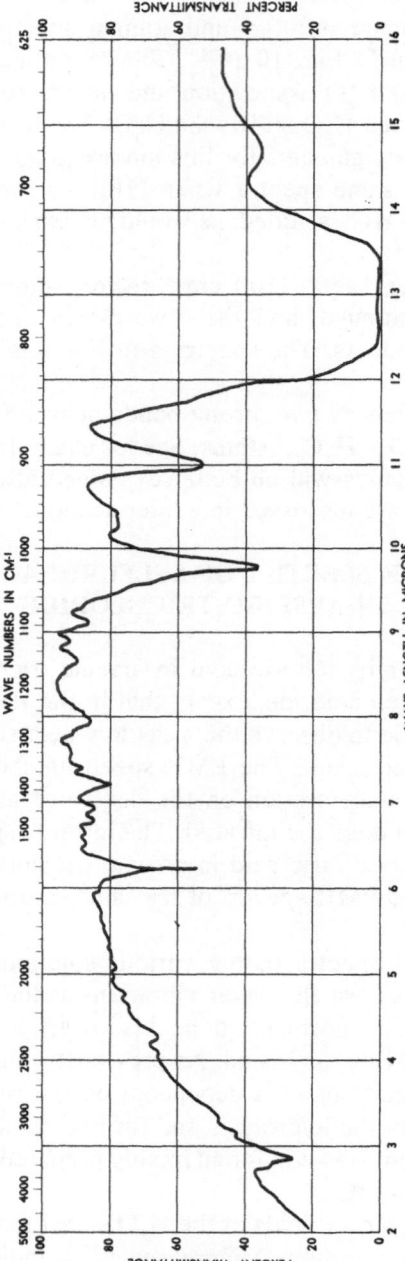

Fig. 11. Fairly dry AsCl₃ prepared by shaking halide with anhydrous MgSO₄. Cell path length 0.5 mm.

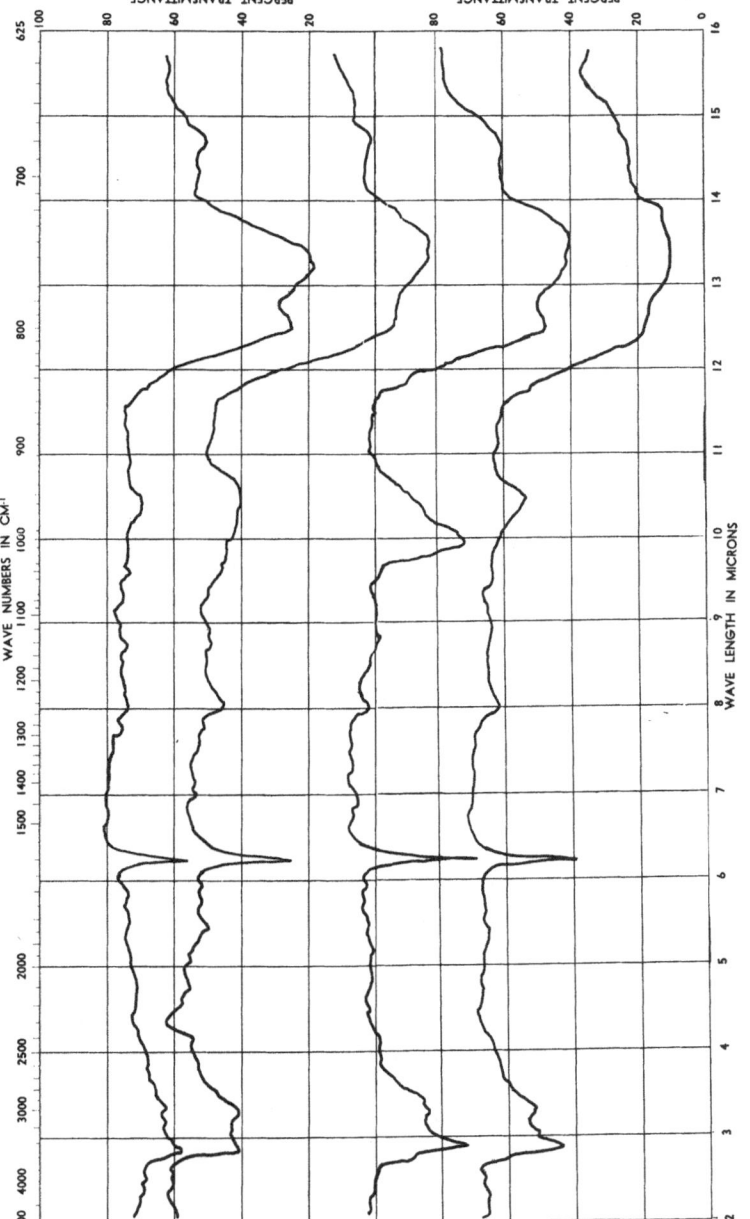

Fig. 12. Infrared spectra of H_2SO_4 in $AsCl_3$. Top curve is 17.8 N H_2SO_4 and the following in order are 8.90 N; 4.45 N; 2.225 N.

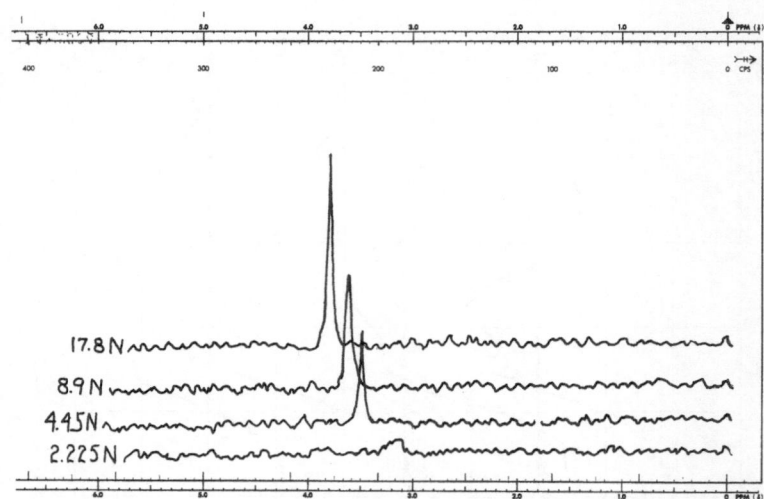

Fig. 13. PMR spectra of H_2SO_4 in $AsCl_3$. (Concentrations as in Fig. 12.)

field side of the H_2O signal such as is found for HCl. However, sulfuric acid is not as soluble as HCl and it is possible that a weak signal exists that cannot be identified.

THE SPECTRA OF CARBOXYLIC ACIDS
IN ARSENIC TRICHLORIDE

In the dimer state formic acid has a band near 917 cm⁻¹ which is the out-of-plane OH bend [5]. In the monomer state this vibration is found at 636 cm⁻¹. The spectra in Fig. 14 are of formic acid liquid and of the acid dissolved in $AsCl_3$. The liquid-state spectrum shows a poorly defined band near 917 cm⁻¹ (top spectrum of Fig. 14). In wet $AsCl_3$ a better defined spectrum of the acid is obtained (middle spectrum of the figure). The most distinct spectrum is obtained if the $AsCl_3$ is fairly dry, and such a spectrum for formic acid is shown as the bottom spectrum of Fig. 14. Here, the out-of-plane OH bend is shifted from the 917 cm⁻¹ position, appearing nearer 910 cm⁻¹. We shall examine this shifting in further detail with other carboxylic acids.

In Fig. 15 the liquid-state spectrum of *n*-valeric acid is presented, and Fig. 16 gives the spectra of two solutions of this acid in $AsCl_3$. The upper curve of the latter figure has more water present, as indicated by the intensity of the 3550 and 1600 cm⁻¹ bands.

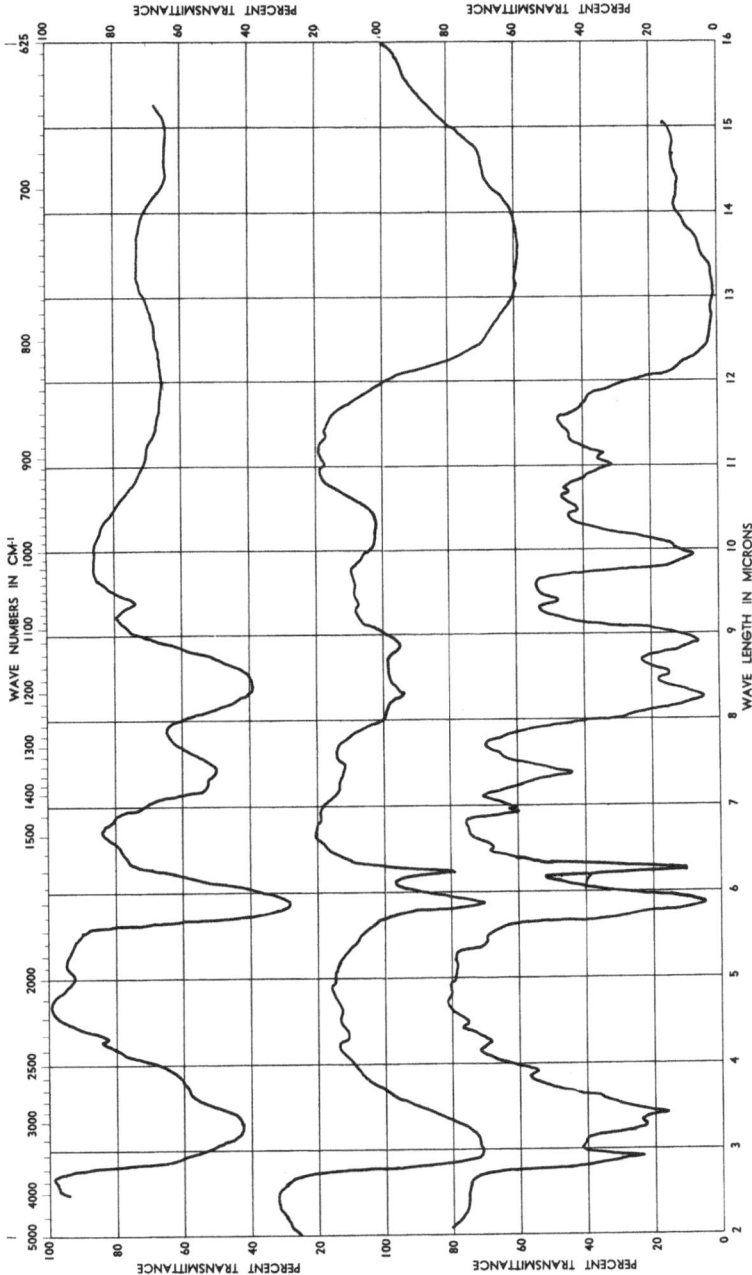

Fig. 14. Infrared spectra of formic acid; top curve is liquid formic acid; middle curve is acid in very wet AsCl₃; bottom curve is acid in fairly dry AsCl₃.

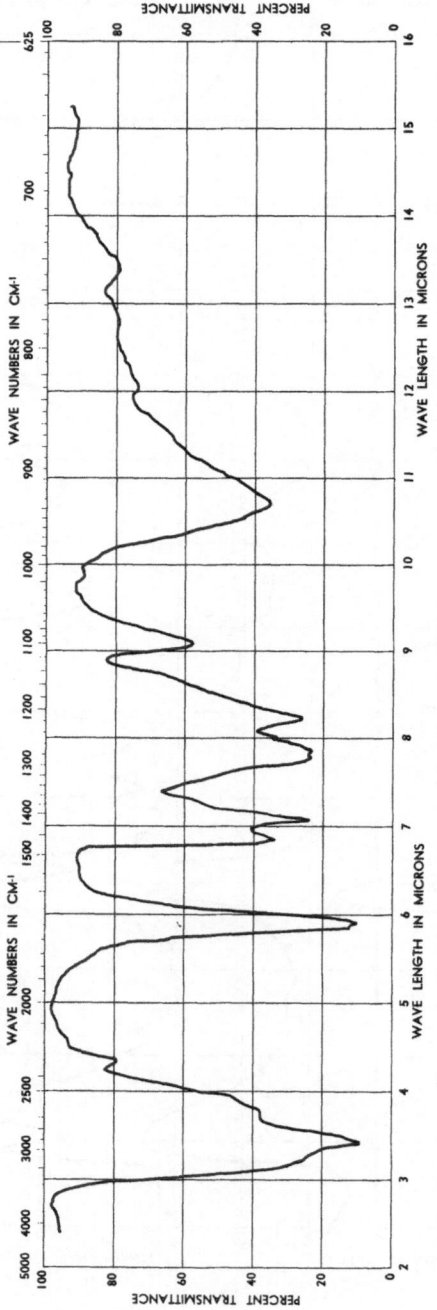

Fig. 15. Infrared spectrum of *n*-valeric acid run as a liquid smear.

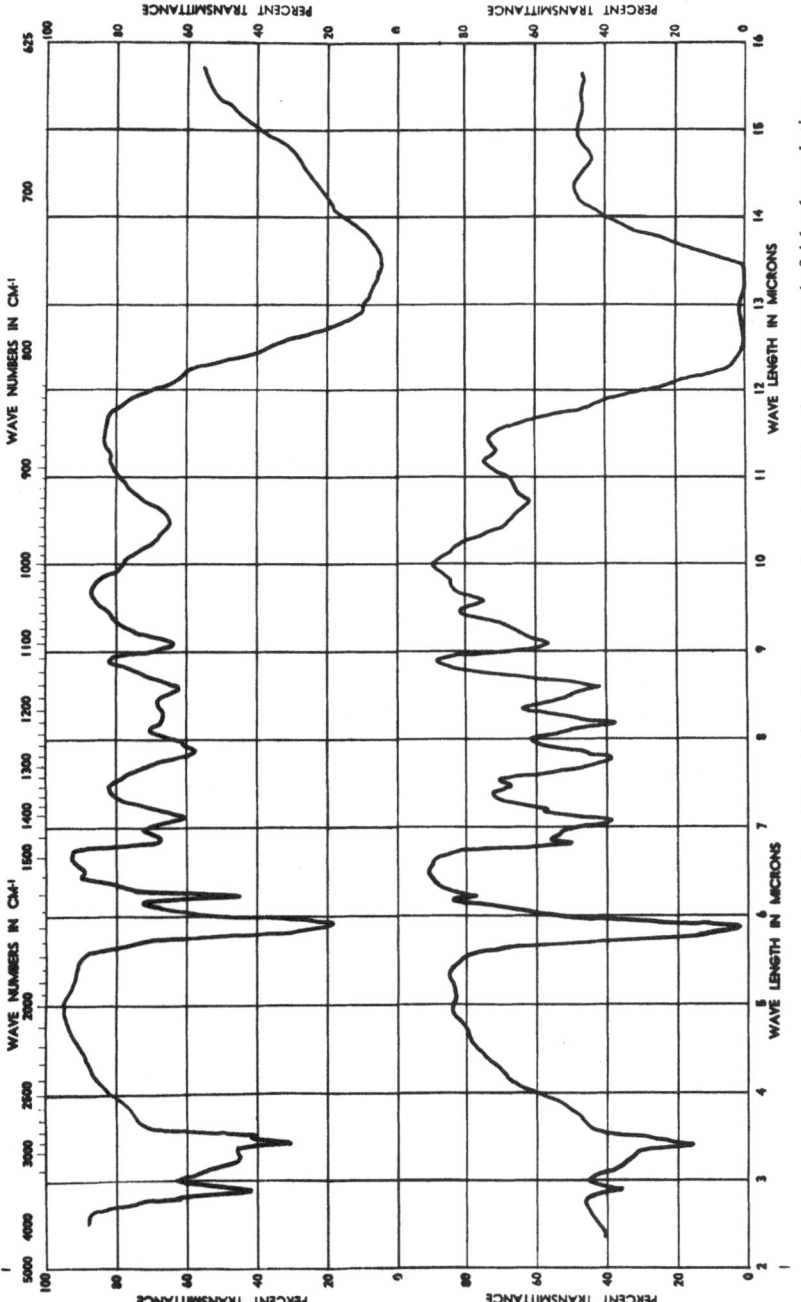

Fig. 16. Infrared spectra of *n*-valeric acid in AsCl₃. Upper curve is wet solution, lower curve is fairly dry solution.

Both of the solutions of *n*-valeric acid are 5% in the halide. The OH out-of-plane dimer band is at 930 cm^{-1} in the liquid, while in the arsenic trichloride the band shifts, depending on the water concentration. In the dry AsCl$_3$ the band is near 930 cm,$^{-1}$ while for the wet AsCl$_3$ it is near 955 cm^{-1}. A band near 1160 cm^{-1} which appears in all carboxylic acid spectra when the acid is dissolved in AsCl$_3$ appears here for the AsCl$_3$ solutions. No simple assignment is obvious for this band. The CH stretch can easily be seen in the AsCl$_3$ solutions.

In Figs. 17 and 18 the spectra for the butyric acid in the liquid state and in AsCl$_3$ solutions are presented. It can be noted that in the AsCl$_3$ solution the 910 cm^{-1} band is again shifted from the liquid-state position. The 1160 cm^{-1} band also is present in these AsCl$_3$ solutions.

INFRARED SPECTRA OF ORGANOPHOSPHORIC ACIDS

While traces of water are difficult to remove from arsenic trichloride, simple distillations can give a fairly dry solvent. When the water content is low the transmission of arsenic trichloride is very high in the $2-12$ μ region even for path lengths as great as 0.7 mm. Adding water when a covalent substance is dissolved in arsenic trichloride reduces the transmission considerably. An example of this is illustrated in the spectra shown in Figs. 19–25. In Figs. 21, 22, 23, 24, and 25 determinations were made in a 0.1-mm cell while in Figs. 19 and 20 very thin smears were used. The compound *bis*-(2-ethylhexyl) phosphoric acid is insoluble in water and is a strongly associated liquid. Its spectrum is shown in Fig. 19. A 33% solution of it in dry arsenic trichloride is presented in Fig. 20. The two spectra are exact duplicates except for the increase in sharpness of the bands for the solution spectrum. The same solution in a cell of 0.1-mm thickness is almost completely opaque and even a 12.5% solution is almost completely absorbing, as shown in Fig. 21. In Fig. 22 a 7.7% solution is shown which will be compared to a 5% solution in wet arsenic trichloride, shown in Fig. 25.

The spectrum of the acid at 3.7% concentration in arsenic trichloride is shown in Fig. 23. Here a 0.1-mm cell is quite satisfactory, indicating that the high absorption of the previous solutions was due to the acid and not to the halide. At this cell thickness weak water bands are seen in the 3600 and 1600 cm^{-1} regions, indicating

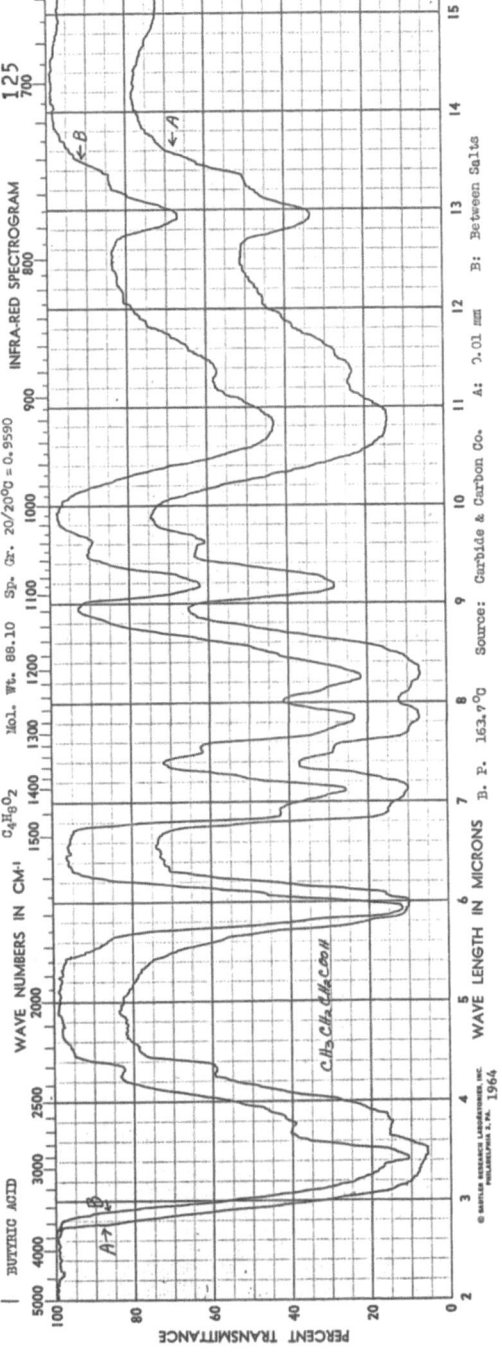

Fig. 17. Liquid-state spectra of butyric acid.

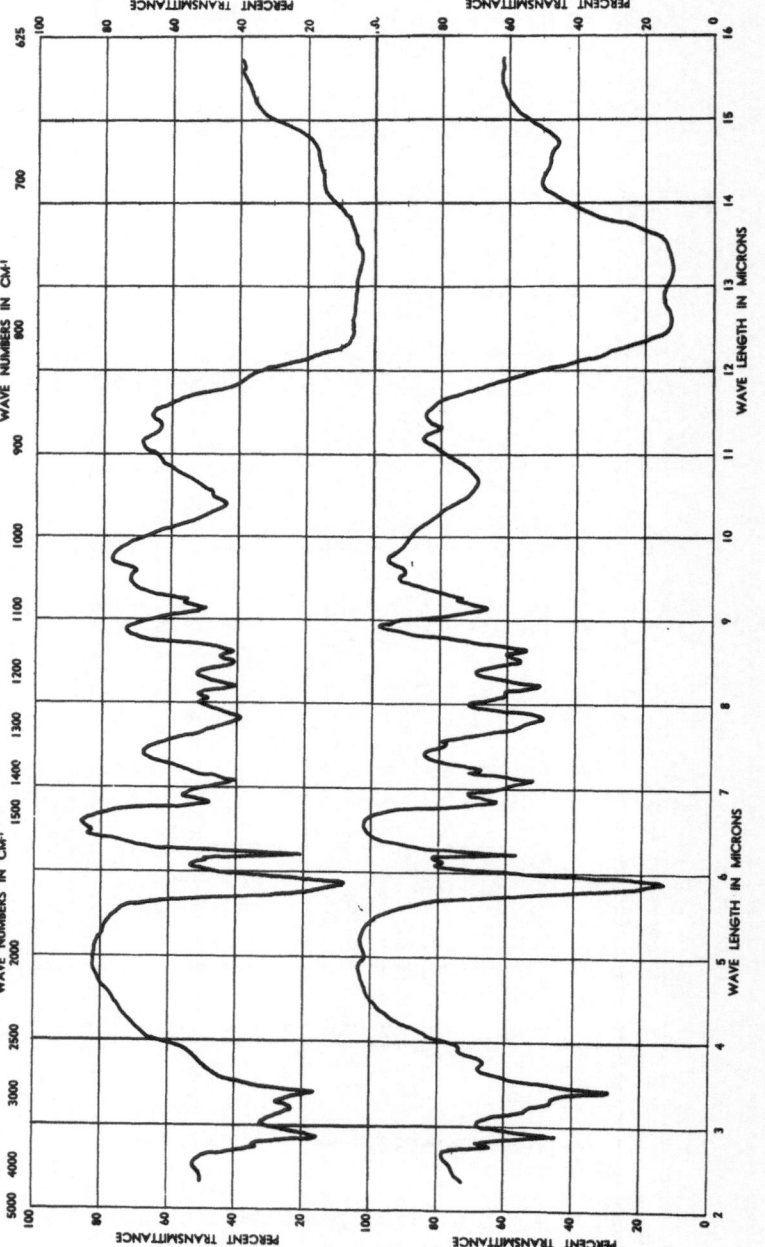

Fig. 18. Infrared spectra of butyric acid in AsCl₃. Upper curve is wet solution, lower curve is fairly dry solution.

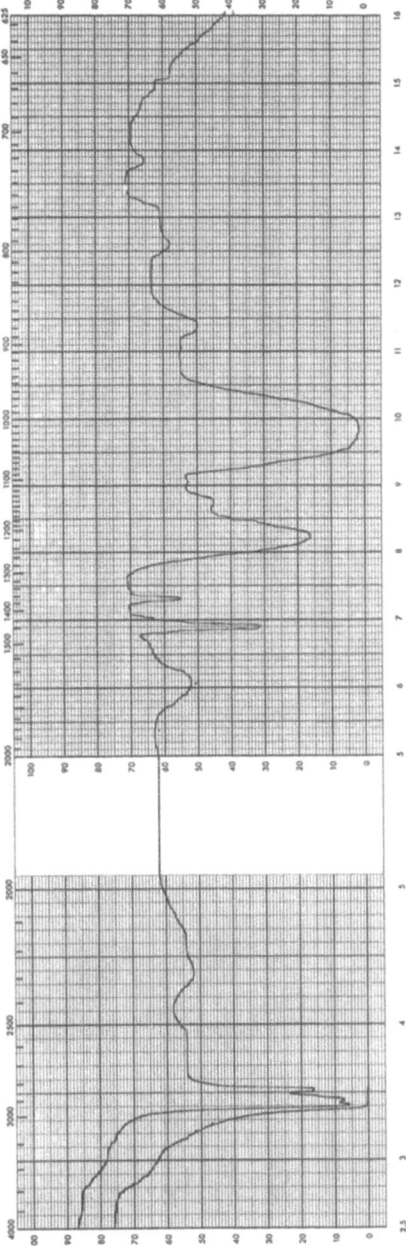

Fig. 19. Spectrum of a liquid-state smear of *bis*-(2-ethylhexyl) phosphoric acid.

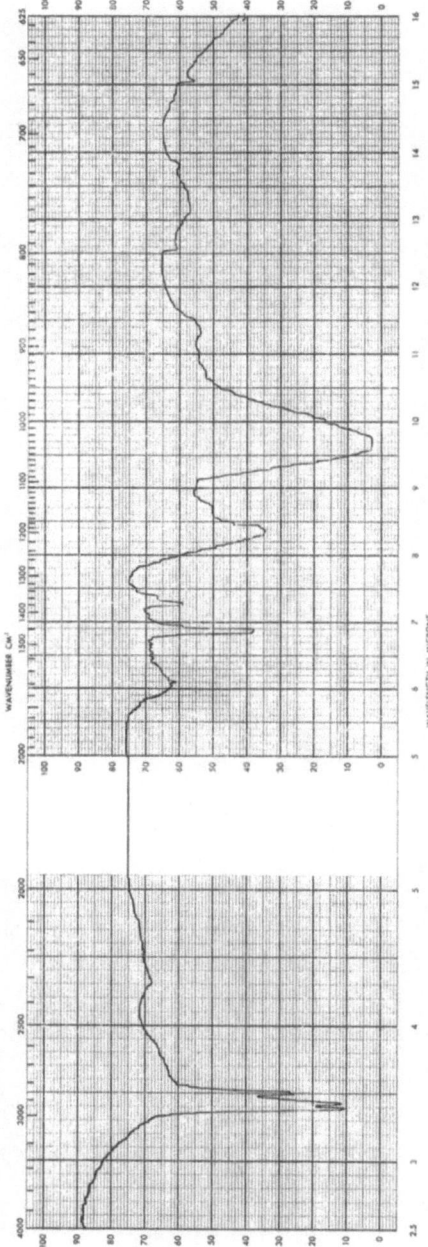

Fig. 20. 33% solution of *bis*-(2-ethylhexyl) phosphoric acid in AsCl$_3$. Spectrum run as a capillary film.

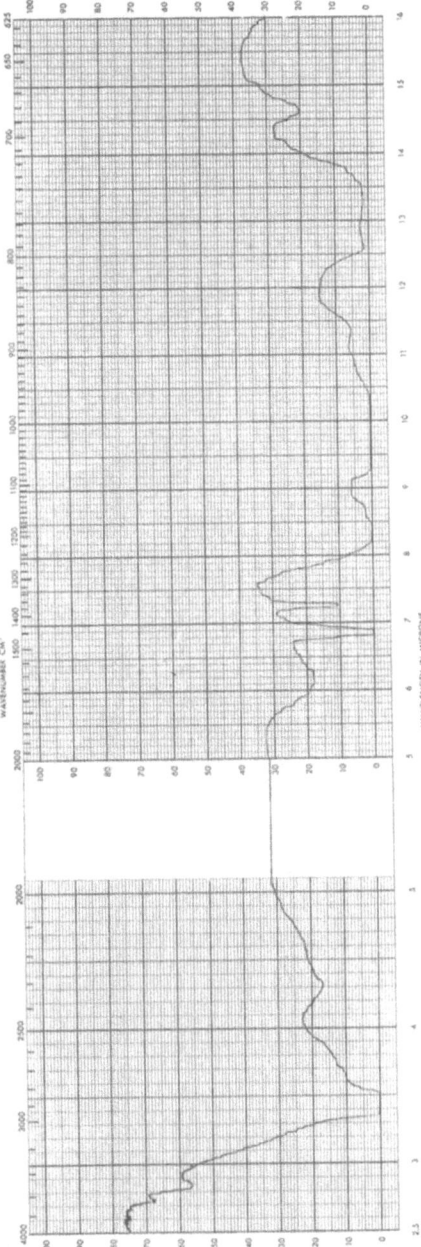

Fig. 21. 12.5% solution of *bis*-(2-ethylhexyl) phosphoric acid in AsCl₃. Spectrum run in a 0.1 mm cell.

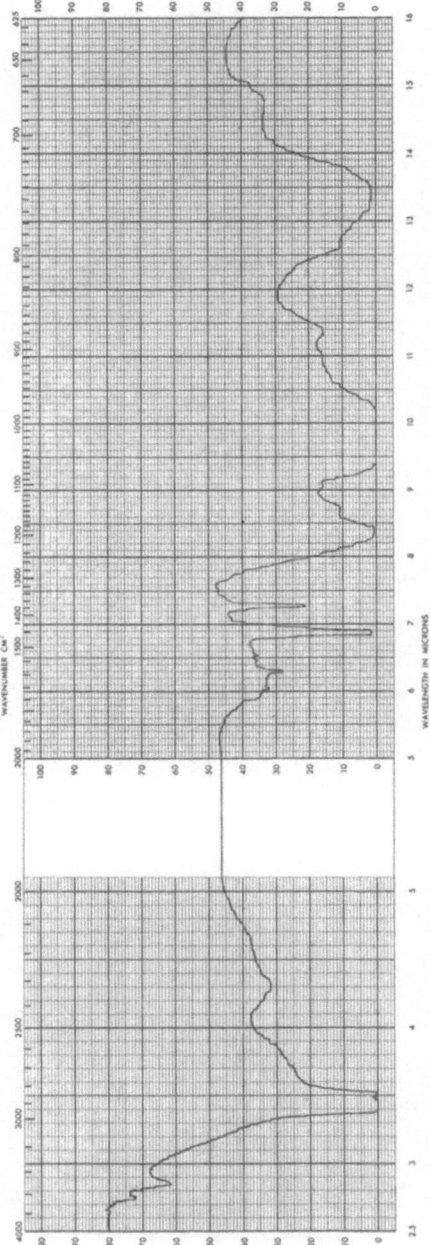

Fig. 22. 7.7% solution of *bis*-(2 ethylhexyl) phosphoric acid in AsCl₃. Spectrum run in a 0.1 mm cell.

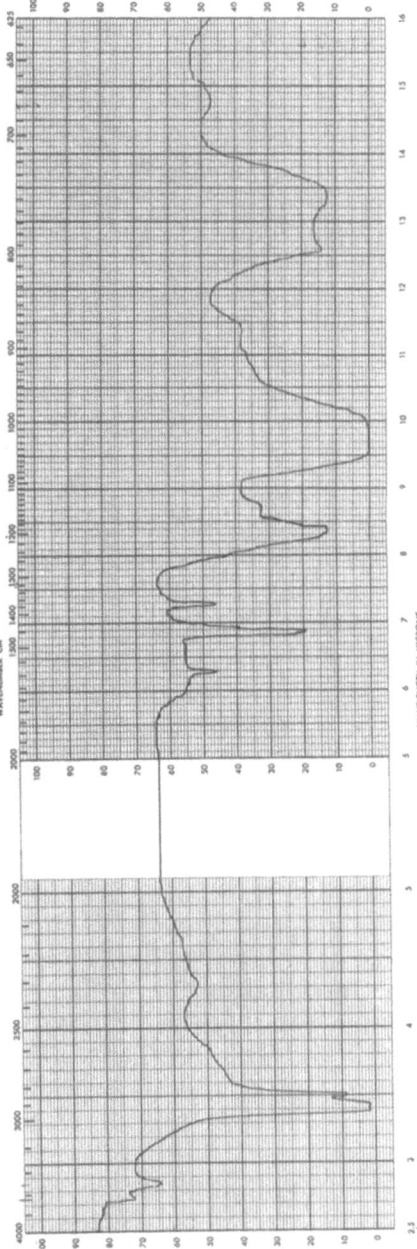

Fig. 23. 3.7% solution of *bis*-(2-ethylhexyl) phosphoric acid in AsCl₃. Spectrum run in a 0.1 mm cell.

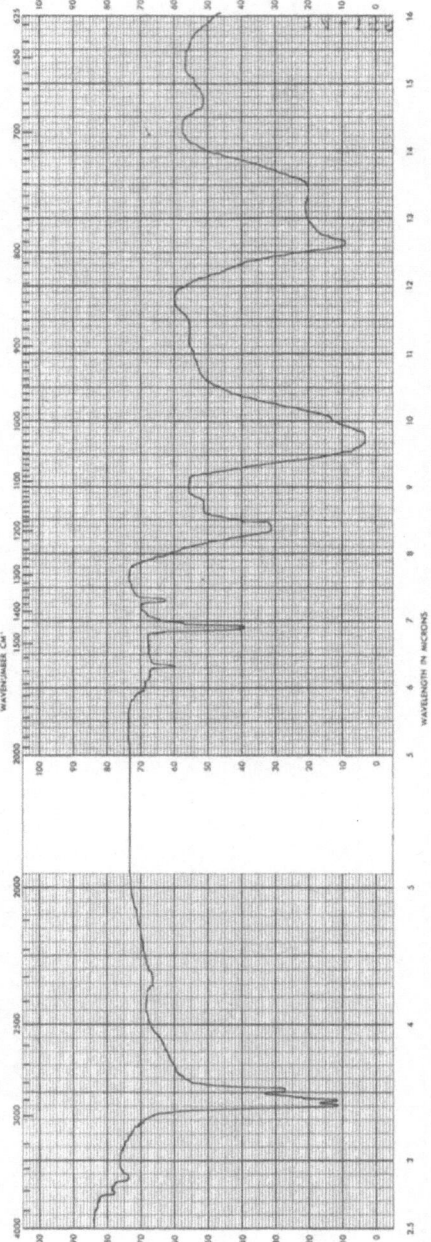

Fig. 24. 2.2% solution of *bis*-(2-ethylhexyl) phosphoric acid in AsCl$_3$. Spectrum run in a 0.1 mm cell.

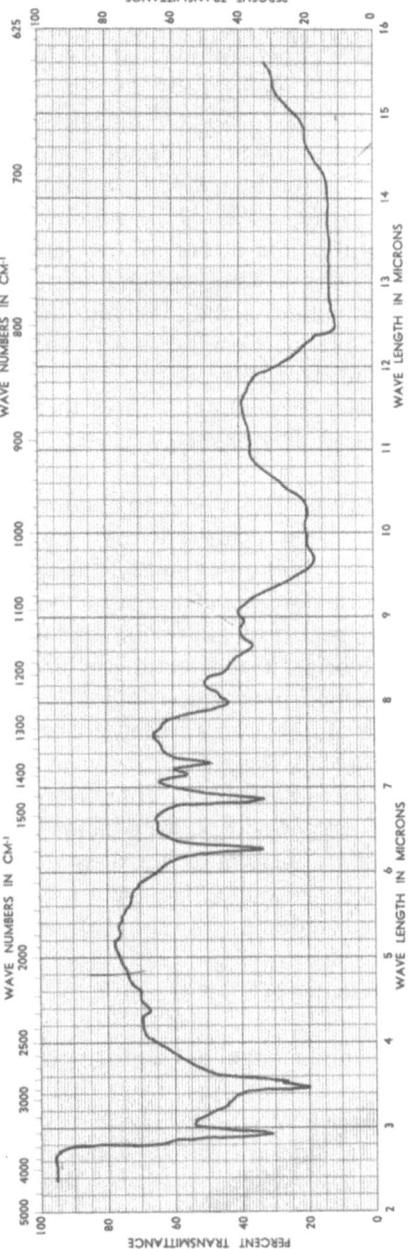

Fig. 25. Infrared spectrum of a 5% solution in AsCl$_3$ of *bis*-(2-ethylhexyl) phosphoric acid.

that the solution has traces of water. In Fig. 24 the most satisfactory spectrum for this acid at this cell path is obtained, the concentration being 2%. At this concentration the acid is partially dissociated, as indicated by the shifting of the 1200 cm⁻¹ band and the sharpness of the 1030 cm⁻¹ band, as seen when the spectrum of the pure acid (Fig. 19) is compared to the spectrum of Fig. 24. These bands are assigned to the P=O stretch and POC bending, respectively.

The spectrum in Fig. 25 is of the acid at 5% concentration in wet arsenic trichloride. The strong water bands are evident. The 1200 cm⁻¹ band is much weaker and shifted here, indicating strong interaction with the water–arsenic trichloride solution. The 1000 cm⁻¹ band is quite broad, partially due to the water–arsenic trichloride bands found here. We consider spectra where associated water bands appear in a solution such as this to be quite unique, since the senior author has never observed this type of spectrum for water except in the vapor state.

The ideal spectrum of this acid is obtained in dry arsenic trichloride. Solid phosphoric acids also give excellent PMR spectra in arsenic trichloride, while other suitable solvents cannot often be found and their solid-state spectra are not very satisfactory. We shall discuss other spectra of materials insoluble in most solvents except arsenic trichloride in the sections which follow. The sharpness of spectra obtained in arsenic trichloride is one of the advantages of using it as an infrared solvent. In addition, the ability of arsenic trichloride to dissociate strongly hydrogen-bonded substances has many advantages. Dr. John Ferraro has studied the infrared spectra of organophosphorous acids and has shown that these acids are mostly dimers in CCl_4 except at very low concentrations. We have compared our PMR spectra with his and find that dissociation occurs at much higher concentrations in $AsCl_3$.

ORGANIC COMPOUNDS IN AsCl₃

Organic compounds such as acetone give sharp spectra in $AsCl_3$. In Fig. 26 spectra of acetone are presented. It should be noted that the carbonyl vibration is at 1696 cm⁻¹, which is found neither for the liquid state nor for the vapor state of acetone. This suggests that an interaction has occurred between the halide and acetone.

The weak CH stretching frequencies of acetone may be due to

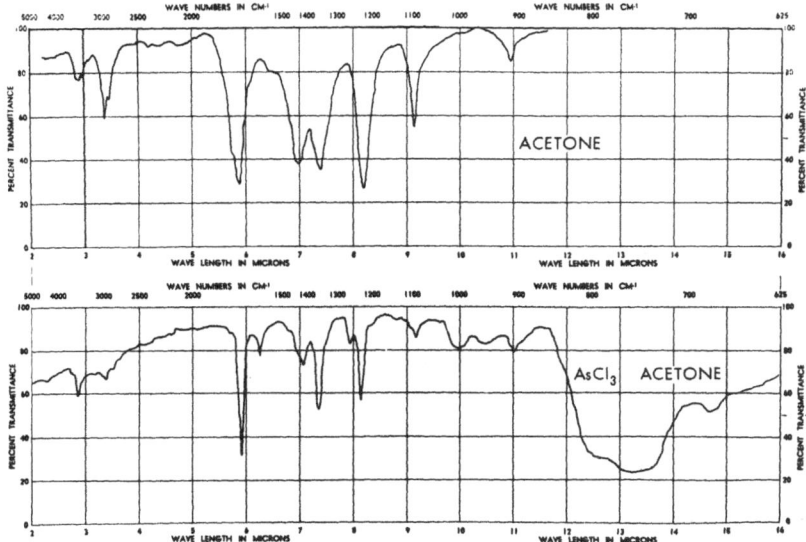

Fig. 26. Infrared spectra of acetone. Upper curve is of the liquid state; bottom curve is of acetone in AsCl$_3$.

the fact that they are sharp bands and the instrument resolution is only fair in this region, giving a low average band intensity.

PMR STUDIES IN AsCl$_3$

The PMR spectra of various materials in AsCl$_3$ support the conclusions drawn from the infrared spectra. For example, that hydrogen-bonded substances are dissociated in AsCl$_3$ is illustrated by the PMR spectra of ethanol in the halide, as shown in Fig. 27. The position of the OH signal in this spectrum is between those of of the CH$_2$ and CH$_3$ groups, as expected for partial dissociation.

Mineral acids in AsCl$_3$ give two signals, one strong signal which appears between that found for liquid water and that of water vapor, and a second weak signal to the low-field side of the liquid-water signal. Both signals are concentration-dependent, as expected for a system where dissociation is occurring. As an example, Fig. 28 gives the PMR spectra of two AsCl$_3$ solutions of H$_2$O compared with the spectrum of liquid H$_2$O. The lower curve is liquid H$_2$O and the middle curve is H$_2$O in AsCl$_3$. The top spectrum was ob-

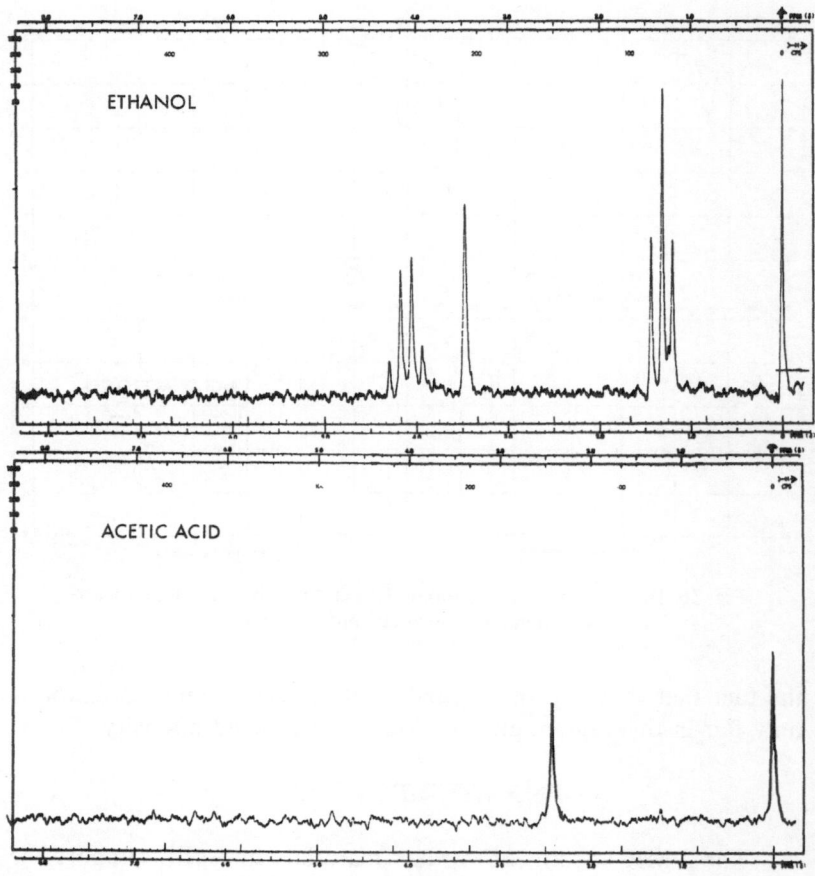

Fig. 27. PMR spectra of ethanol and acetic acid in AsCl$_3$.
(Conditions as marked on curve.)

tained when HCl gas was added to the H$_2$O – AsCl$_3$ solution. The
gas causes further dissociation of the water present and the strong
signal of the spectrum which is due to H$_2$O moves to higher fields.
The solution is more acidic and the weak signal moves to lower
fields.

The same general type of PMR spectra found for mineral acids
is obtained for carboxylic and organophosphorous acids, and Figs.
29 and 30 are typical. In each case an acidic OH signal is found
which is concentration-dependent and moves to high fields as concentration of the acid is decreased. This is seen for the signal near

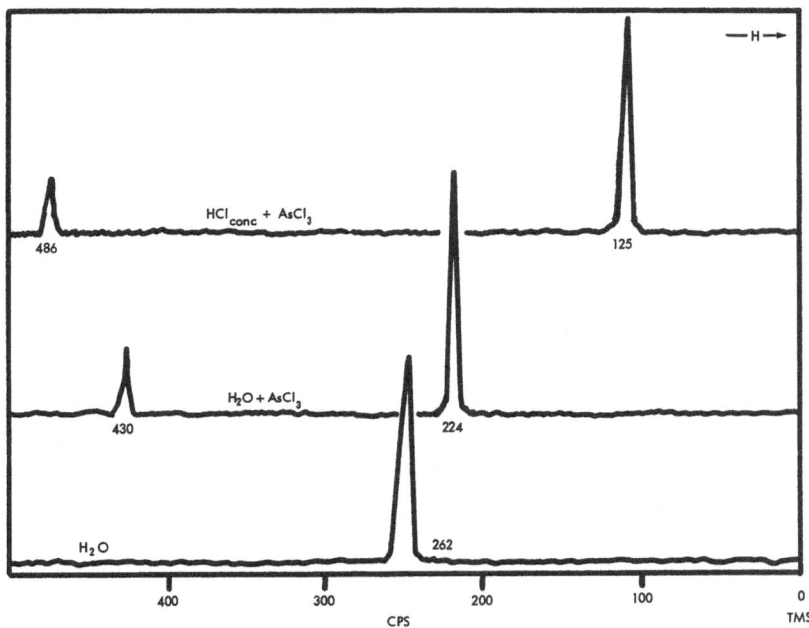

Fig. 28. The PMR spectra of water. Lower curve is that of the liquid and upper curves are of water and HCl in AsCl₃.

6.0 ppm for liquid *n*-valeric acid in Fig. 29. The relative positions of the nonshifting signals of the CH_2 and CH_3 groups also differ for the liquid state and AsCl₃ solutions, as expected when a solvent with the magnetic susceptibility of AsCl₃ is used.

The solvent should have no protons if PMR spectra are to be measured. AsCl₃ fulfills this requirement and also has strong solvent properties. This makes it a useful solvent for polymers which normally give poor spectra because of high intrinsic viscosity.

We have studied polyamides, polyimides, polyesters, polyvinyl chlorides, and phenolics in AsCl₃. As an illustration, the spectra of a series of model compounds and a polymer having the structural units present in the compound are shown in Fig. 31. The sharpness of the lines is indicative of the good solvent properties of the halide i.e., the viscosity of the polymer has been reduced to where the spectrum has sharp lines.

It is of interest in the model compound of the top spectrum that both the CH_2 and CH_3 groups give two kinds of signal, indicative of

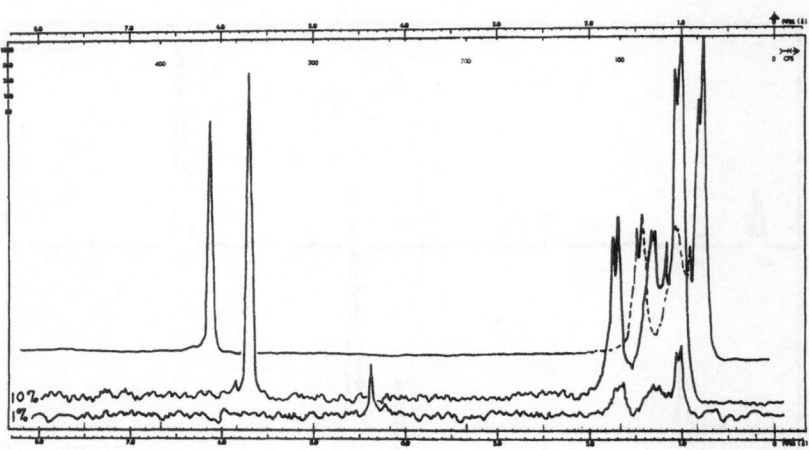

Fig. 29. PMR spectra of *n*-valeric acid. Lower curve is 1% solution in AsCl₃;
middle curve is a 10% solution; top curve is the pure liquid.

the two kinds of environment they are present in. It is possible to
trace the signals of each type of environment the structural groups
are in, and assign the same group in the polymer shown as the bottom curve.

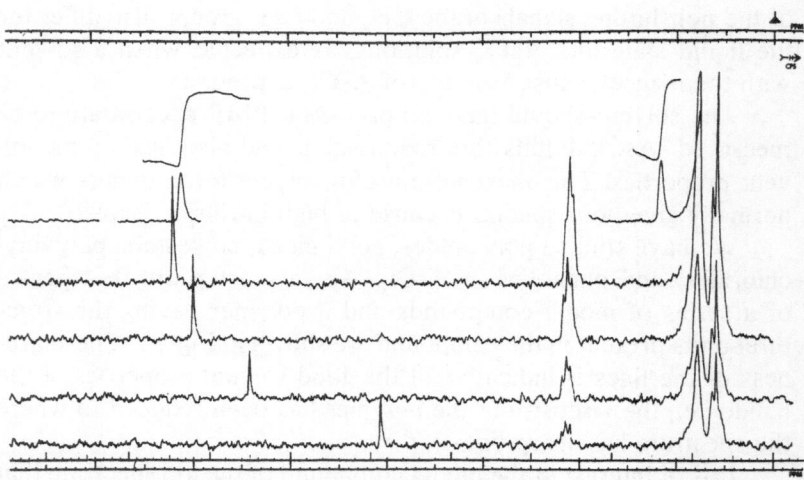

Fig. 30. PMR spectra of *bis*-(2-ethylhexyl) phosphoric acid.

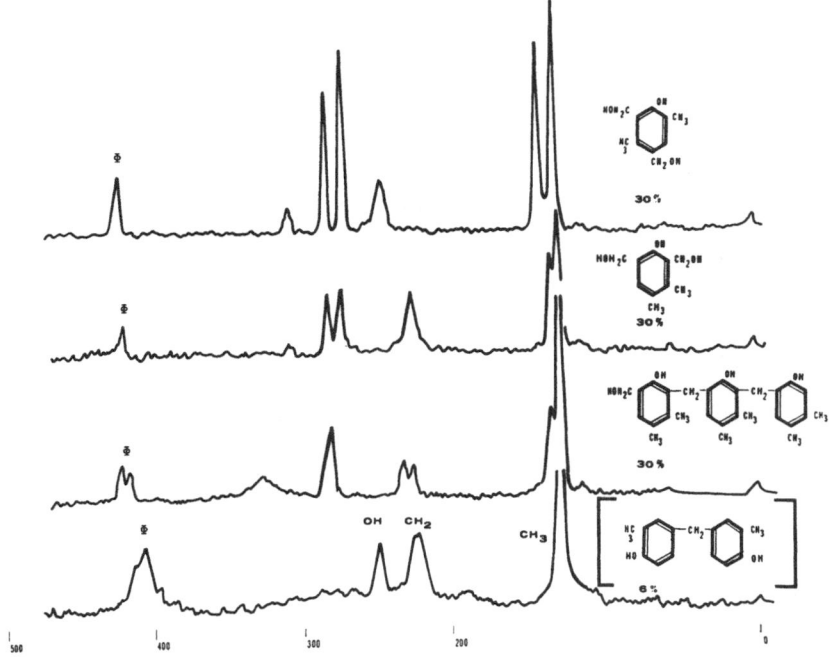

Fig. 31. PMR spectra of a series of model compounds and of a polymer (bottom curve) having structural units from these model compounds.

MIXED SOLVENT

A solvent similar to AsCl₃ is SbCl₃, although the latter is solid at room temperature. At 80°C (melting point of SbCl₃) polymers as difficult to dissolve as polyimides will dissolve in liquid SbCl₃.

We have found that SbCl₃ will dissolve in AsCl₃ in equimolar mixtures giving a liquid at room temperature. The resulting solution has solvent properties which are similar to those reported above for AsCl₃, and in addition will dissolve materials not soluble in AsCl₃, such as polyimides. We are investigating this new solvent system.

ACIDITY OF H₂O–AsCL₃ SOLUTIONS

We have treated H₂O–AsCl₃ solutions as highly acidic media. To illustrate how similar the solutions are to acidic media we have examined trimethylamine hydrochloride in wet AsCl₃. It has been reported [6, 7] that in water solutions below pH = 4 a doublet for the methyl group will be obtained in the PMR spectrum of the hydrochloride. Above pH = 4 only a single line is observed. We have obtained the doublet with the expected cps separation of the two lines of the doublet, indicating solutions are at least below pH = 4.

REFERENCES

1. Progress in Inorganic Chemistry, Vol. 2, Cotton, F. A. (ed.), Interscience, New York, (1960), p. 63.
2. Gutman, V., Z. *anorg. u. allgem. Chem.* 266:331 (1951).
3. Giguere, P. A., and Falk, M., *Can. J. Chem.* 35:1195 (1957).
4. Mullhaupt, J. T., and Hornig, D. F., *J. Chem. Phys.* 24:169 (1956).
5. Pimentel, G. C., McClellan, A. L., The Hydrogen Bond, W. H. Freeman Company, New York, (1960), p. 131.
6. Grunwald, E., Loewenstein, A., and Meiboom, S., *J. Chem. Phys.* 27:630 (1957).
7. Grunwald, E., Loewenstein, A., and Meiboom, S., *J. Chem. Phys.* 27:641 (1957).

Infrared Spectroscopy of Biological Materials

Leopold May

Department of Chemistry
The Catholic University of America
Washington, D.C.

Infrared spectroscopy may be used to provide valuable information in the study of biological systems. It can assist in the study of the structure of various macromolecular components within tissue. It can also supplement other chemical or physical methods of analysis for the qualitative or quantitative determination of various components within tissue. In specialized cases, infrared spectroscopy affords rapid qualitative and quantitative identification of constituents, for example, in the analysis of gallstones and lipides.

Sampling

The applicability of infrared spectroscopy to biological systems depends upon the nature of the sample that is involved in the analysis. For example, we can divide the samples used into several different types: exhaled vapor (gas techniques); body fluids (solution techniques); tissues (solid techniques); macromolecular components (both solid and solution techniques); and minor constituents such as metabolites (both solid and solution techniques). The sampling also depends upon whether we are going to analyze our material directly or to extract from the material some component we are interested in studying. The type of sampling further depends upon the purpose of the spectroscopic examination: qualitative

213

analysis, quantitative analysis, or structural determination of a component in the material.

Gases. For the analysis of gaseous samples such as exhaled air, including carbon dioxide, metabolic products, and other components such as carbon monoxide, ethyl alcohol, etc., the sampling is essentially that which is used in ordinary gas work. For obtaining samples from an animal or a patient, one technique is to have the patient blow a pillow-sized plastic bag full of his exhaled air and then to transfer this gas sample to a 10-m cell [1]. In this way one can analyze for carbon monoxide, ethyl alcohol, etc., in the blood stream. It is possible also to monitor for carbon dioxide in exhaled lung air by analysis of the 4.26μ absorption band using an infrared photometer.

Of great importance to biochemists and others studying the human organism is the analysis of blood flow and variations in it due to different physiological conditions. For example, nitrous oxide is used a great deal in studying blood flow. Here again infrared analysis could be utilized to give results more easily than the manometric methods which have been used previously. However, results are generally 6% higher with infrared techniques than with manometric techniques [2].

Liquids. In general the milieu of biological samples is water. For example, most tissues contain between 65 and 85% water (bone and teeth contain less, 10–30%). Living tissue therefore is essentially water, so that it is desirable to study the spectra of most biological materials in aqueous solution, since this is the natural medium of these materials.

The spectrum of water is shown in Fig. 1. Although it has large absorption due to the stretching and deformation of the OH bonds, it is possible to compensate for these bands. This is done either by putting a sample of water in the reference beam or, in the case when one is dealing with tissue, by using a salt solution that more naturally represents the millieu in which tissue is found. Usable spectra can be obtained when a screen is placed in the reference beam. The compensated spectrum is also shown in Fig. 1. In many cases it is not possible to obtain a window in which an absorption band can be observed, in such cases heavy water may be used. The absorption spectrum of heavy water is shown in Fig. 2. The absorption bands of heavy water are compensated for in the same manner as are those of ordinary water (Fig. 2). It should be

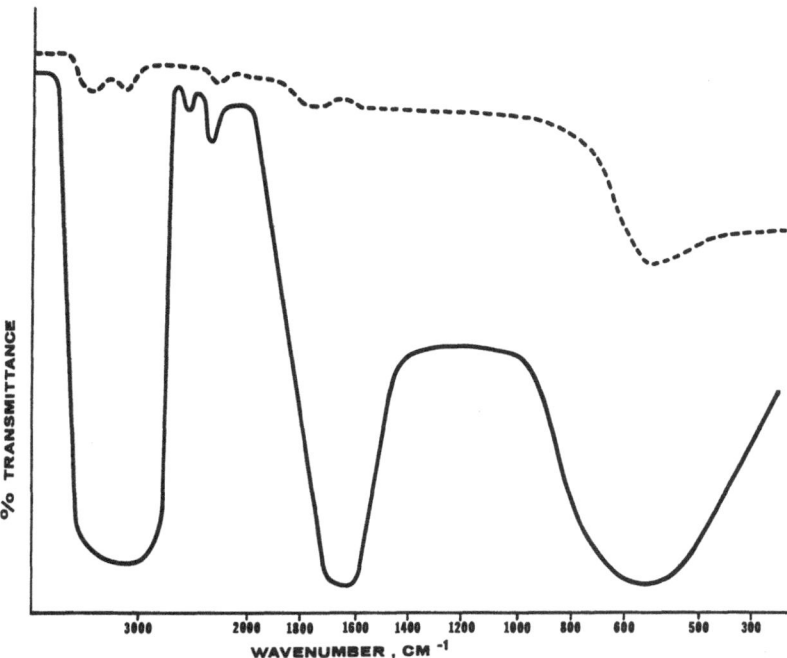

Fig. 1. Spectrum of water. Uncompensated ————; compensated —————.

pointed out that heavy water sometimes alters the structure of components. For example, proteins are known to be denatured in this medium, with the irreversible breaking of many hydrogen bonds taking place.

The use of water determines the kind of cell material that should be used. Table I presents the various cell materials that can be used with water. It should be pointed out that silver chloride tends to corrode the metal backing plates and is more difficult to use. Irtran-2 can be used quite readily at different pH's. Although sodium chloride is quite soluble in water, it has been found that saturated solutions of sodium chloride can be used quite easily. Polyethylene bags have been used with a great deal of success, but it should be noted that polyethylene has absorption bands at approximately the same position as does Nujol. Cells of the polyethylene bags have been constructed using thin metal spaces.

If the material to be examined, e.g., metabolites or drugs, can be extracted by organic solvents, there is an easy solution to the

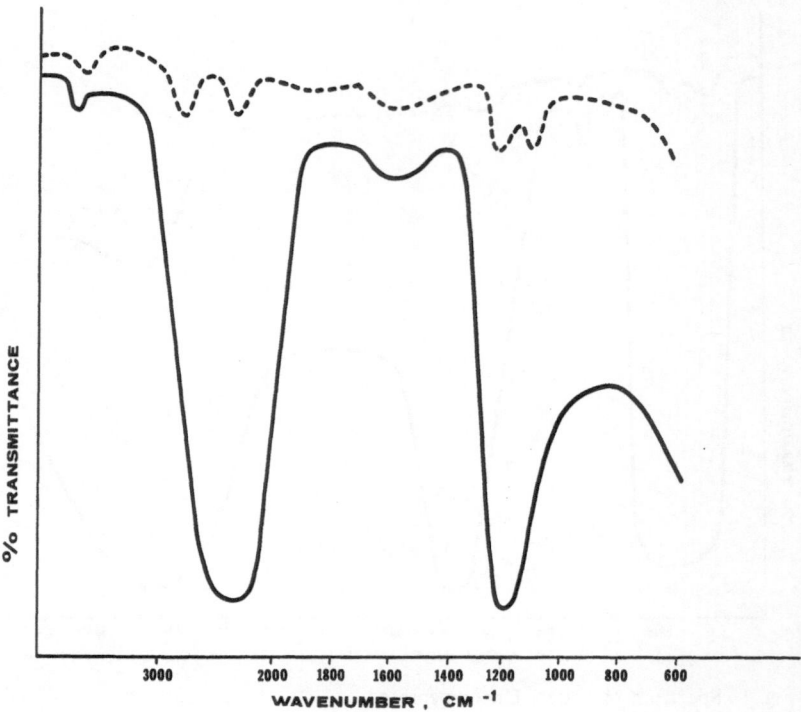

Fig. 2. Spectrum of D_2O. Uncompensated ——— ; compensated — — — — —.

problem. In this case the usual solution techniques can be used or the material can be examined as a solid suspended in KBr, Nujol mull, or even in polyethylene disks.

Tissue Sampling

Biological materials are in general rather complex mixtures. It is desirable to separate or fractionate the tissue to remove the components in which one is interested. Some fractionation techniques that have been used successfully are: for analysis of lipides in blood, chromatography; for analysis of various blood protein fractions, alcoholic precipitation; and for analysis of proteins in tissue, centrifugation. One other method of reducing the complexity of cellular material is to fractionate tissue by centrifugation into nuclear, mitochondrial, and microsomal fractions. In this case, one then has to separate the materials in which one is interested from these primary separations. However, this does provide a means of reducing the complexity of the material.

In general, it would be desirable from a morphological point of view to examine a tissue sample directly itself. The section of tissue can be placed on a silver chloride plate. It requires, however, some means of fixing these samples. Chemical fixation such as the use of formalin tends to remove nucleic acid. In the procedure of fixation, a dehydration step using xylene and alcohol removes some lipides. It has been found desirable in many situations to eliminate this type of fixation. Freeze-drying of the tissue eliminates many of the difficulties. In this case a section of the tissue is cut frozen and placed on a cold silver chloride plate, which is kept on dry ice. It has been found necessary to warm the silver chloride plate very slightly so the section will adhere to the plate. This is done most easily with the finger. The plate is then placed *in vacuo* while cold, and the water is removed by vacuum distillation. It has been found that a high vacuum is not desirable because the sample tends to be flaky.

With bacteria or blood material, it is only necessary to prepare a smear, which is a suspension of these materials in water. This can be placed directly on silver chloride plate and evaporated *in vacuo* to remove water. One can also use Irtran-2 plates or plates of the other materials listed in Table I.

Another approach to this situation is to homogenize the tissue.

TABLE I
Cell Materials for Use with Aqueous Solutions

Material	Thickness, mm	Spectral range, cm^{-1}
Glass	5.5	>3700
Fused quartz	5.9	>3000
Sapphire	2.6	>2500
LiF	5.4	>1800
CaF_2	5.34	>1100
BaF_2	4.68	> 900
AsS_3	1.95	> 900
Ge	5.82	5000 – 700
NaCl[a]		5000 – 700
Irtran-2	3.0	> 600
AgCl	2.0	> 400
KRS-5	2.0	> 250
Polyethylene[b]		> 250

[a]Use saturated NaCl solution.
[b]Strong absorption bands at 2700 – 2900, 1450, 1370, and 700 $cm.^{-1}$

This eliminates the problem of fixation but destroys the morphology. In this case the homogenate can be prepared as a film on silver chloride or any other suitable material (Table I). One can also lyophilize and prepare films from a homogenate of the dried tissue.

One problem found with tissue samples is scattering. Since most of these samples are nonhomogeneous in terms of the distribution of the particle, broad bands are found. One solution, of course, is to use a material such as Nujol to reduce scattering. The bands of the Nujol are then found in the sample. Other suspending agents such as perfluorokerosine have been used in the regions where the Nujol absorbs. A second method is to compare the sample and the blank at about 10μ or where there is no absorption. The blank is then ground until the absorption at this wavelength is about 95%. This will reduce or compensate for the scattering that one would find in the sample itself. A third method is to adjust the reference beam with an iris diaphragm until about 95% transmittance is found in the region where no absorption is observed (10μ). This has been found to be effective with bacterial smears.

Another useful method is to suspend the solid in a solid medium. This has the advantage of reducing scattering if one uses potassium bromide or another alkali halide. The sample can then be saved for further analysis by other methods such as ultraviolet spectroscopy and chemical analysis. The process consists of mixing the sample with the alkali halide and pressing this mixture in a die after evacuation. It has been found that the use of pressure may alter the structure of the suspended sample. Since most of the alkali halides are hygroscopic, one cannot readily eliminate the band at 3μ due to OH stretching. The choice of alkali halide depends upon the region to be examined. Potassium bromide disks can be used to 30μ, potassium chloride to 35μ, and cesium bromide to 40μ. The exact pressure used varies with the halide. The sharpness of many absorption bands in tissue increases when one compares tissue suspended in KBr with the original tissue section. Experiments with enzymes show that some are denatured when disks are made.

It has been suggested that agar or gelatin films should be used in examining the infrared spectra of proteins [3]. The sample is suspended in an agar suspension, which is then placed in a glass cylinder seating on a glass plate. The surface of this glass plate has been previously treated with silicone oil. The assembly is then placed at 35–37°C for about 20 hr to remove the water, and the resulting film is removed from the glass plate with a razor blade. This has the advantage that the protein is suspended in a medium similar to itself, but it also has a great disadvantage in that both gelatin and agar are protein materials with absorption bands similar to those found in proteins. One uses an agar film with no protein in the reference beam. To compensate for the absorption of the agar, a blank of the agar film is used. It is possible that the heating to 37°C will alter the structure of the samples, and experiments with enzymes showed this to be true.

Another sampling technique which we may use for studying metabolites and low-molecular-weight compounds involves the use of polyethylene disks [4]. The sample is mixed with 50 mg of low-melting polyethylene (Microthene, U.S. Industrial Chemicals) and placed between two cover slips. The mixture is melted in a melting point apparatus at about 100–120°C. The two cover slips are pressed together with cork and then allowed to cool. A translucent disk is obtained after the cover slips are removed. This preparation has the following advantages: there is a decrease of scattering as would be found in halide disks, it is nonhygroscopic, and it is transparent beyond 15μ with enough windows in the sodium chloride region to make it useful.

A comparison of some of these preparative methods has been made using enzymes [5]. Three preparations examined were evaporated films on silver chloride, agar films, and KBr disks. Although tissue shows the same spectrum in all these preparations, it is possible that proteins may be denatured during the preparation for spectral examination. The spectra of one enzyme, alcohol dehydrogenase, indicated that it was in the β or extended form in all preparations. Enzymatic analyses confirmed that the enzyme preparations were denatured. When these experiments were repeated with enzyme ribonuclease, the spectra indicated that the enzyme was in the helical form. However, enzymatic analysis showed that the enzyme was denatured to a great extent in the agar film, but not in the other preparations. It would appear that infrared structural analyses of proteins and macromolecular cellular components should not be made without some independent measurement of the denaturation of these components. It would be preferable to use water solutions when studying these materials. If there is no interest in the structure of these components, but merely in the posi-

TABLE II
Components Found in Tissues

Major Components	
	Macromolecular components
	Unconjugated
	Proteins
	Nucleic acids
	Polysaccharides
	Mucolipides
	Conjugated
	Nucleoproteins
	Lipoproteins
	Glycoproteins
	Micromolecular components
	Lipides
	Peptides
	Nucleotides
	Nucleosides
	Disaccharides
Minor Components	
	Molecular components
	Amino acids
	Sugar
	Organic acids, citric acid, etc.
	Organic bases, purine, etc.
	Steroids
	Hormones
	Porphyrins
	Minerals
	PO_4^{3-}, Na^+, Ca^{2+}, CO_3^{2-}, K^+, etc.

tion of the bands not influenced by the structure, any convenient method of preparation will give this information.

Biological Systems

In general, the problem of analysis of tissue depends upon the material present and the analyte being sought. Table II lists some of the components that one would find. The tissue itself is dominated by the spectrum of the macromolecular components, which are present to the largest amount. If living tissue is being examined, it is dominated by the water spectrum. In the analysis of tissue, we can consider three types: organic tissues such as kidney, liver, heart, and brain; inorganic – mineralized tissues such as bones and calculi;

and body fluids—blood, urine, saliva, bile and spinal fluid.

The objective of the analysis of these tissues would be an attempt to determine what variation has occurred in the spectrum of the normal tissue. What is the effect of some physiological situations such as disease, drug action, psychological change, and what changes might occur in the normal composition with aging? It might be of interest to analyze for a particular component which changes in response to some physiological stress. For example, in animal experiments or clinical research, we might analyze for the appearance of some particular component or the change in the amount of this material. In some instances our interest would be in determining the structural variation in some macromolecular component as we impose some variation in the physiological situation. It is valuable in the diagnosis of certain diseases to determine whether a particular component has changed in its concentration or has even been altered in its structure. Finally, infrared spectra of tissues may be utilized to determine whether there have been changes in the amount or structure of various components as part of the study of the mechanisms of the reactions within the cell.

Clinical Analysis

Clinical infrared spectroscopy has been used with materials that are easily obtained: body fluids (blood, urine, and bile) and mineralized excrements (calculi and gallstones). Some examples of these analyses are discussed to illustrate the usefulness of the infrared technique in clinical analysis.

Calculi and Gallstones. Chihara and his co-workers have studied urinary calculi and gallstones for a number of years [6]. Sampling was done immediately after removal from the patient, and within 20–30 min. an analysis of the type of stone could be made. Previously, it required about one day for chemical analysis. They found that Nujol mulls are better if gallstones were analyzed, but potassium bromide disks were satisfactory for calculi. The major components of these materials can be divided into inorganic and organic. One can quite easily distinguish between inorganic and organic stones by simply looking at the spectra. Generally, the inorganic samples have fewer bands than the organic stones, and by an analysis of these bands, one can readily determine which components are present. For trace analyses of elements such as iron and magnesium, emission-spectrographic methods must be used.

Blood. Blood is essentially a mixture of water, protein, inorganic salts such as phosphate, and small amounts of low-molecular-weight organic compounds. The blood can be examined as smears or directly in a liquid cell. Blood smears are simply prepared by placing a drop of the blood on a silver chloride plate and evaporating to dryness. The blood serum can be examined directly by using the usual techniques for water solutions. It requires wide slits and differential techniques using an expanded scale to observe variations in the absorption. Erley and his co-workers [1, 7] have examined a large number of spectra and have found that generally the spectra of sera from individuals with different diseases are very similar. The variation seems to appear most pronouncedly at the 7.35μ band. This band appears to vary from individual to individual as well as daily for a single individual. From absorbance data of the spectra of normal individuals, 36 nonredundant band ratios were calculated in the region of $7-10\mu$. They found that the same results could be observed simply by looking at the spectra when comparing the spectra. To determine what was responsible for the variation in the 7.35μ region, various isolated fractions of blood (albumin, alpha globulin, beta globulin, and gamma globulin) were examined. It appears that the variation of this 7.35μ band is due to variation in the relative amounts of the different proteins in the serum. There are several other variations that must be considered. The cells (BaF_2) must be reset each time, so that a thickness correction must be made. One also finds variations in baseline, and the usual photometric errors must be considered.

Lipide Analysis

It is quite important in the investigation of the course and causes of many diseases to analyze for lipides in blood and tissues. The infrared spectra of extracts from blood and tissue provide a rapid method for the analysis of the various lipides and components. The general procedure is first to extract the lipides from blood or tissue using lipide solvents such as a chloroform–methanol mixture. It has been found that the infrared spectra of the crude extracts does not give sufficient information to yield complete analyses. Consequently the lipides in the extract are separated using chromatographic columns (silic acid or celite), and the spectra of the various fractions are then used for the complete analysis.

TABLE III
Lipide Analysis

Fraction	Composition	Band, cm^{-1}
I	Cholesteryl esters	1170 or 1725
II	Glycerides (GLY)	1740
	Unesterified fatty acids (FA)	1710
	Cholesterol (CHOL)	1050
III	Phospholipides[a]	1070
	Lecithin	1390, *1750*
	Sphingomyelin	1390, 1540, *1660*
	Cephalin	*970*, 1450, *1750*

[a]Bands listed for individual phospholipides are in addition to the 1070 cm^{-1} band. The bands reported in *italic* numerals were used for analysis.

The eluded fractions from the columns are evaporated to dryness. The resultant solids can either be dissolved in known volumes of carbon disulfide [8] or prepared as potassium bromide disks [9]. In the former method the infrared spectra of the fractions are measured in the region from 900 to 2000 cm^{-1}, and the composition of each fraction and bands used are given in Table III. The absorbances of the various bands are measured in the usual manner. Since fraction I contains only cholesteryl esters, it is only necessary to measure the 1725 cm^{-1} band and directly calculate the concentration of the cholesteryl esters. The band at 1070 cm^{-1} in fraction III represents the phospholipides, and it is assumed that it is mostly lecithin. However, if the fraction is later found to include cephalin and sphingomyelin, a correction would have to be made. The analytical bands for these are also given in Table III. It requires the use of additional bands and the use of equations similar to those which are used for determining the composition of fraction II. To analyze fraction II, the absorbances of three bands are determined, since this fraction contains three components, each of which contributes to the absorption of the bands listed. The equations for the absorbance at each one of these bands are

$$A_{1740} = A_{GLY}^{1740} C_{GLY} + A_{FA}^{1740} C_{FA} + A_{CHOL}^{1740} C_{CHOL} \tag{1}$$

$$A_{1710} = A_{GLY}^{1710} C_{GLY} + A_{FA}^{1710} C_{FA} + A_{CHOL}^{1710} C_{CHOL} \tag{2}$$

$$A_{1050} = A_{GLY}^{1050} C_{GLY} + A_{FA}^{1050} C_{FA} + A_{CHOL}^{1050} C_{CHOL} \tag{3}$$

Since cholesterol absorbs very slightly at 1710 and 1740 cm^{-1}, equations (1) and (2) reduce to

$$A_{1740} = 0.100\ C_{GLY} + 0.016\ C_{FA}$$
$$A_{1710} = 0.009\ C_{GLY} + 0.166\ C_{FA}$$

where numerical values are given for the absorptivity indices [8]. Solution of these two equations gives the concentrations of the glycerides and total unesterified fatty acids. Equation (3) becomes

$$A_{1050(CORR)} = A_{1050(MEAS)} - 0.01\ C_{GLY} - 0.006\ C_{FA} = A_{CHOL}\ C_{CHOL},$$

and the concentration of cholesterol can be determined directly.

In the second method [9], the eluted samples are incorporated into KBr disks by a two-step freeze-drying process. First, a 10% solution of KBr is lyophilized. Then, 50 mg of the freshly prepared KBr is mixed with an organic solution of the lipide sample and lyophilized. The disk is prepared immediately by the usual procedure. The disk is weighed, and the infrared spectrum measured using a beam condenser system. The amount of the lipide is determined by comparison with standards prepared by the same procedure.

BIBLIOGRAPHY

Review Articles and Books

A. General

"Biological Applications of Infrared Spectroscopy," Bauman, R. P. (ed.), *Ann. N.Y. Acad. Sci.* 69:1-254 (1957).

Clark, C., "Infrared Spectrophotometry," in: Physical Techniques in Biological Research, Vol. I, Oster, G., and Pollister, A. W. (eds.). Academic Press, New York, 1956, pp. 206–325.

Fraser, R. D. B., "The Infrared Spectra of Biologically Important Molecules," Progr. in Biophys. and Biophys. Chem. 3:47-60 (1953).

Freeman N. K., "Infrared Spectrometry," *Advan. in Biol. and Med. Phys.* 4:167-238 (1956).

Wood, D. L., "Infrared Spectrometry," in Methods in Enzymology, Vol. IV, Colowick, S. P., and Kaplan, N. O. (eds.), Academic Press, New York, 1957.

B. Bibliographies

Clark, C., and Chianta, M., "A Bibliography of Infrared Spectra of Biochemicals," *Ann. N.Y. Acad. Sci.* 69:205-253 (1957).

Levine, C., "A Bibliography on Applications of Infrared Spectrophotometry in the Biological Sciences, Environmental Health Center, U.S.P.H.S., Cincinnati, Ohio.

C. Tissues

May, L. and Grenell, R. G., "Infrared Spectral Studies of Tissues," *Ann. N.Y. Acad. Sci.* 69:171-189 (1957).

Norris, K. P., "Infrared and Its Application to Microbiology," *J. Hyg.* 57:326 (1959).

Norris, K. P., "Infrared Spectra of Micro-organisms," *Advan. Spectr.* 2:293-330 (1961).

D. Proteins, Peptides, Amino Acids

Doty, P., and Geiduschek, E. P., "Optical Properties of Proteins," in: The Proteins, Vol. I (A), Neurath, H., and Bailey, K. (eds.), Academic Press, New York, 1953, pp. 393–460.

Elliott, A., "The Infrared Spectra of Polymers," *Advan. Spectr.* 1:214-87 (1959)

Greenstein, J. P., and Winitz, M., Chemistry of the Amino Acids, Vol. 2, John Wiley and Sons, New York, 1961, pp. 1695–1716.

Sutherland, G. B. B. M., "Infrared Analysis of the Structure of Amino Acids, Polypeptides, and Proteins," *Advan. Protein Chem.* 7:291-318 (1952).

E. Nucleic Acids and Components

Beaven, G. H., Holiday, E. R., and Johnson, E. A. "Optical Properties of Nucleic Acid," in: The Nucleic Acid, Vol. I, Chargaff, E., and Davidson, J. N. (eds.), Academic Press, New York, 1955 pp. 545–51.

F. Lipides, Fats

Deuel, H. J. Jr., The Lipids, Vol. I, Interscience, New York, 1957, p. 77.

Holman, R. T., Lundberg, W. O., and Malkin, T., "Infrared Absorption Spectroscopy in Fats and Oils," *Progr. in Chem. Fats Lipids* 2:292 (1954).

O'Connor, R. T., "Application of Infrared Spectrophotometry to Fatty Acid Derivaties," *J. Amer. Oil Chem.* Soc. 33:1-15 (1956).

O'Connor, R. T., "Recent Progress in the Application of Infrared Absorption Spectroscopy to Lipid Chemistry," *Ibid.* 38:648-59 (1961).

Schwarz, H. P., "Infrared Absorption Analysis of Tissue Constituents Particularly Tissue Lipides," *Adv. Clin. Chem.* 3:1-34 (1960).

G. Carbohydrates

Barker, S. A., Bourne, E. J., and Whiffen, D. H., "Use of Infrared Analysis in the Determination of Carbohydrate Structure," *Methods Biochem. Anal.* 3:213-45 (1956).

H. Steroids, Hormones, and Vitamins

Dobriner, K., Katzenellenbogen, E. R., and Jones, R. N., Infrared Absorption Spectra of Steroids, An Atlas, Interscience, New York, 1953.

Rosenkrantz, H., "Analysis of Steroids by Infrared Spectrometry," *Methods Biochem. Anal.* 2:1-55 (1955).

Rosenkrantz, H., "The Infrared Analysis of Vitamins, Hormones, and Coenzymes," *Ibid.* 5:407-51 (1957).

Cole, A. R. H., "Infrared Spectra of Natural Products," *Fortschr. Chem. org. Naturstoffe* 13:1-69 (1956)—Steroids, Terpenes, Cartenoids.

REFERENCES

1. Erley, D. S., "Infrared Analysis in Biomedical Research," 14th Annual Mid-America Spectroscopy Symposium, May 1963.
2. Hafkenschiel, J. H., *Federation Proc.* 14:67 (1955).
3. Androsine, J., and Krupa, C., *Biochem. J.* 335:212 (1961).
4. May, L., and Schwing, K. J., *Appl. Spectry.* 17:166 (1963).
5. May, L., and Boccalatte, R., *Anal. Biochem.* 9:1 (1964).
6. Chihara, C., *et al., Chem. Pharm. Bull.* 7:622 (1959); 8:771 (1960).
7. Stewart, R. D., Skelly, N. E., and Erley, D. S., *J. Lab. Clin. Med.* 56:391 (1960).
8. Nelson, G. J., and Freeman, N. K., *J. Biol. Chem.* 234:1375 (1959).
9. Schwarz, H. P., *Adv. Clin. Chem.* 3:1 (1960).

Near-Infrared Spectroscopy:
Theory, Instrumentation, and Applications

James D. McCallum

Beckman Instruments, Inc.
Fullerton, California

INTRODUCTION

The spectral region extending from the top of the visible region ($0.78\ \mu$) up to and overlapping the beginning of the rock-salt region ($3\ \mu$) has been termed the near-infrared region. It is desirable to segregate this region from both the infrared and the ultraviolet and visible regions for several reasons. First, for optimum performance when moving from the infrared down into the near infrared it is necessary to change instrument optics, sources, detectors, and sample-handling techniques. Second, absorption bands observed in the near infrared are due to vibrational transitions rather than to electronic transitions as are those observed in the visible and ultraviolet regions.

The near infrared has also been referred to as the overtone region. This designation has sound theoretical foundation, since absorption bands observed in the near infrared are due primarily to overtones and combinations of hydrogenic (CH, OH, NH) stretching and bending fundamentals which occur in the rock-salt region.

The characteristic and essentially constant positions of hydrogenic stretching modes makes the near-infrared region quite valuable for quantitative, qualitative, and structure studies on molecules containing CH, OH, and NH groups. In addition to this, the significant environmental effects produced by concentration, solvent interaction, and functionality of some of these groups, notably the OH group, make the near infrared particularly useful for studies of hydrogen bonding and other forms of inter and intramolecular associations.

The purpose of this discussion is to outline briefly the theory governing near-infrared absorption, the factors governing the position and intensity of overtones and combination modes, the essential features of near-infrared instrumentation, various sample-handling techniques, and some of the analytical and theoretical problems to which near-infrared spectroscopy has been applied.

BASIC THEORY OF NEAR-INFRARED ABSORPTION

We will not delve deeply into the theory of absorption since this is the subject of a separate and quite extensive discussion [1]. Let it suffice to say that to a first approximation most vibrating molecules behave as harmonic oscillators. That is, the frequency of an absorption band characteristic of a given group is a function of the masses of the atoms and the strength of the bond between them in the following way:

$$\nu = \frac{1}{2\pi} \sqrt{\frac{k}{M_r}} \tag{1}$$

where ν is the frequency in cm^{-1}, k is the force constant, and M_r is the reduced mass of the vibrating atoms, equal to $m_1 m_2/(m_1 + m_2)$. In the case of CH, NH, and OH vibrations, M_r reduces to $m_2/(1 + m_2)$ since $m_1 = 1$ for hydrogen. The value of M_r for these three groups, therefore, is 0.92, 0.93, and 0.94, respectively.

Examination of the expression for a harmonic oscillator thus reveals that it matters little what is attached to the hydrogen, because the value of M_r is controlled by the mass of the hydrogen atom. This factor generally tends to make hydrogenic vibrations extremely reliable as to position. The most influential factor in determining the frequency then is the force constant k. The value of k varies with the nature of m_2 and also with the nature of the bond, i.e., single, double, or triple. The relationship between k and bond order is approximately 1 : 2 : 3 going from single to triple bond.

Therefore, in spite of the fact that M_r values for these various groupings are so close together (a fact which would make it difficult to differentiate among CH, NH, and OH bands) there is enough difference in k to separate the characteristic absorption bands of these groups by an appreciable amount. This fact, coupled with the constancy of hydrogenic group frequencies, is a major contributor to the specificity of near-infrared bands.

Overtones

If vibrating molecular groups behaved as pure harmonic oscillators, there would be no absorption at overtone and combination frequencies. However, most natural vibrations are not truly harmonic.

These deviations from true harmonicity are similar to those observed in a pendulum in that the degree of anharmonicity increases with an increase in the amplitude of the vibration. Since hydrogen stretching vibrations involve very light atoms and, therefore, are relatively large in amplitude, one would expect significant deviations from harmonicity.

If the degree of deviation from true harmonic motion (the anharmonicity) is known, it is possible to calculate the frequencies of the overtone bands. The expression relating anharmonicity to overtone frequency is (to a first approximation)

$$\nu_v = v \, \nu_0 (1 - vx) \tag{2}$$

where ν_v is the observed frequency for the vth harmonic, ν_0 is the fundamental frequency, and x is the anharmonicity constant, which is positive. For CH vibrations x is from 0.01 to 0.05 [2]. For a precise calculation of anharmonicity values, higher terms should be added to this expression, but it is useful in predicting approximately the positions of overtone bands.

Combination Bands

Combination bands are those which are observed at frequencies equal to the sum of and/or difference between two or more fundamental frequencies. Summation bands are commonly observed in the near-infrared spectra of many molecules, but difference tones are rather unusual at room temperature.

Fermi Resonance

Fermi resonance is a phenomenon observed when coupling occurs between two vibrations of the same symmetry when their normal frequencies are very close together [3]. It is usually manifested by a significant change in frequency and intensity if the bands

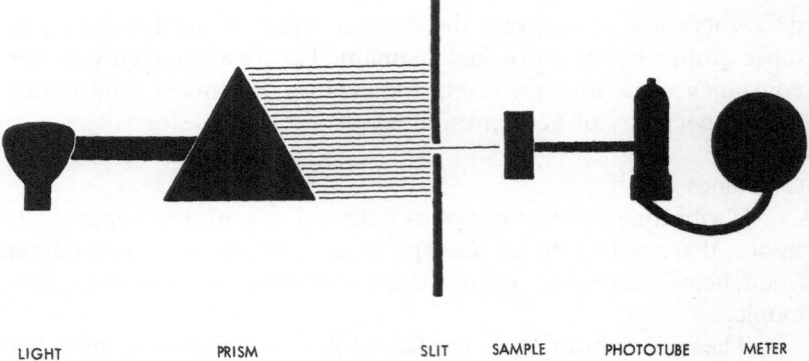

LIGHT   PRISM   SLIT   SAMPLE   PHOTOTUBE   METER

Fig. 1. Basic spectrophotometer components.

are close together, and an increase in the intensity of the weaker band with little change in band position if the bands are not closely spaced. Such resonance is commonly observed in the spectra of methyl halides and is well illustrated by Kaye [3]. An example will be given later.

FUNDAMENTALS OF NEAR-INFRARED INSTRUMENTATION

As mentioned earlier, one of the reasons for classifying the near-infrared as a separate spectral region is the instrumentation employed in this region. It is usually required of near-infrared instrumentation that it bridge the gap between the ultraviolet ($160-360$ mμ) and the rock-salt region ($2-16$ μ). This requirement places on the designer the responsibility of selecting the appropriate components to cover this spectral region in a manner that is as economical as possible in terms of the necessity for changing instrument conditions. In this regard, we are somewhat fortunate in having at our disposal sources which emit well, detectors which are very sensitive, and dispersing elements capable of producing high dispersion.

General Considerations

To begin our discussion of instrumentation, consider that any spectrophotometer, no matter how complicated it may appear externally, is composed of the same five basic components. They are shown schematically in Fig. 1 [4], and consist of a radiant

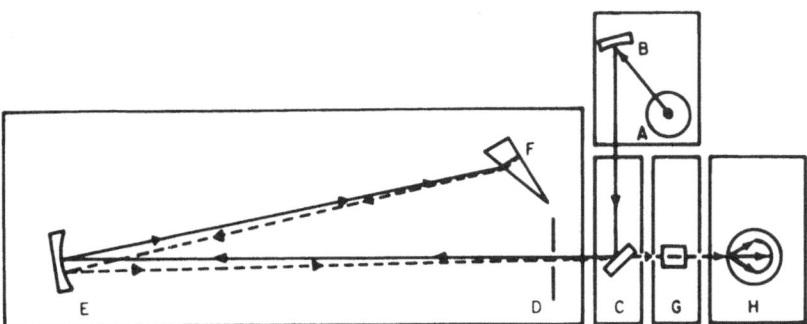

Fig. 2. DU optical diagram.

energy source, a detector, a monochromator, a sample compartment, and a readout mechanism. Figure 2 shows one typical way
in which these might be arranged.

Sources

If we start our discussion with the first link in the chain, the
radiant energy source, we immediately run into some problems.
First, no single source is available which will cover the entire
ultraviolet, visible, and near-infrared regions. Second, sources are
quite nonlinear in their output. Consequently, at some wavelengths
we have energy to spare, while at others energy is limited. As we
shall see later, there are several ways to cope with this situation.

The most commonly used sources for ultraviolet, visible, and
near-infrared spectrophotometry are the hydrogen-discharge lamp
for the ultraviolet and the tungsten lamp for the visible and near
infrared. The ordinary hydrogen lamp is limited on the lower end
(below about 200 mμ) by the absorption of its silica envelope. By
using special window material, this range may be extended to 160
mμ. On the upper end its useful range stops at 360 mμ. Beyond this,
the spectrum of hydrogen changes from a continuum to the Balmer
series of discrete emission lines. The emission of a typical hydrogen
lamp is illustrated in Fig. 3. The tungsten lamp is also limited at
both ends. The normal glass envelope absorbs ultraviolet radiation
below 360 mμ and near-infrared energy at about 3500 mμ. Therefore the tungsten lamp is limited to the visible and near infrared
regions. Its peak energy point is at about 1000–1500 mμ at normal
operating temperature.

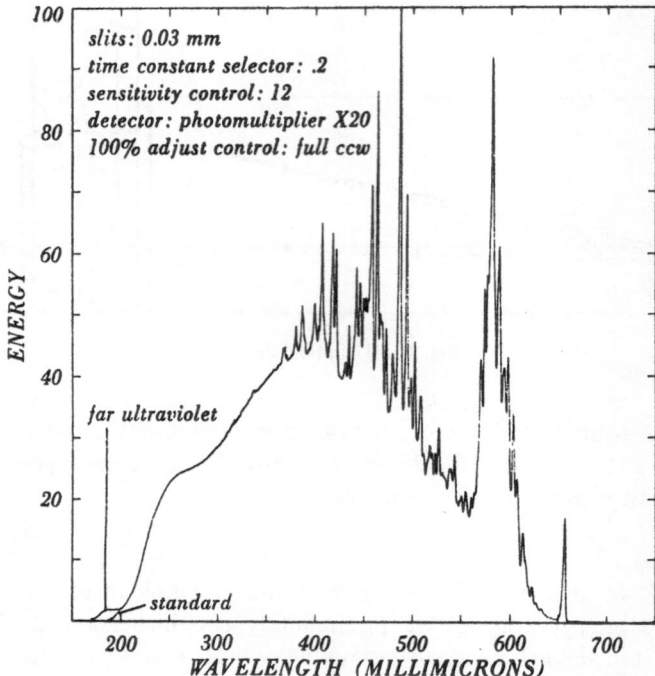

Fig. 3. Hydrogen lamp spectrum.

Source problems were among the first limitations encountered in the design of early spectrophotometers. At the time, convenient UV sources were not available. The first compromise was to use a tungsten light source, and operate the instrument in the visible region, but design the instrument so that the conversion could be made as easily as possible when good ultraviolet sources became available. In time a hydrogen lamp was developed which was suitable in size, power requirements, and energy output.

Many sources have since been used in the ultraviolet, including deuterium, xenon, and mercury lamps, the last of which is commonly used both as a convenient wavelength calibrator and as an energy source, especially for fluorescence studies. The hydrogen lamp is still the universal standard.

In the near infrared region, the most commonly used and most convenient source is the tungsten lamp. Globars and Nernst glowers have been used, but are both somewhat limited in this wavelength region. The necessity for cooling a Globar and regulating current to a Nernst glower increases their cost and decreases their convenience as sources. Arc lamps, such as the carbon arc, may also be used if more intense radiation is needed.

The last source I would like to mention is the flame, in which the sample itself becomes the source and the monochromator serves to analyze the emission or absorption spectrum thus produced.

Detectors

Detectors also represent an area where manufacturers must bow to present technology in making their choices. Detector studies have progressed much over the years, but the choice is still somewhat limited to a relatively few types. These may be divided roughly into three classes: thermal, photoemissive, and photoconductive. Thermal detectors, such as the thermocouple, possess a very wide frequency response range, but are limited somewhat in signal-to-noise ratio by their sensitivity to room radiation. Thermocouples are basically heat detectors and are bathed in an environment of heat under the usual conditions. Pneumatic detectors, such as the Golay cell or the Luft-type pneumatic detector, have a very wide range and, in the case of the Luft cell, may be quite selective by proper choice of the filling gas. Detectors of this selective type are in wide use in nondispersive analyzers. Photoemissive detectors, such as the red- and blue-sensitive phototubes and the photomultiplier, give adequate response from 160 mμ to about 1 μ and are commonly used in many commercial spectrophotometers. Photoconductive devices, such as the lead sulfide (PbS) cell, give excellent response in the visible and near-infrared region and, therefore, have found wide commercial use as detectors. Response curves for a typical lead sulfide cell are shown in Fig. 4.

In spite of the progress which has been made in detectors, if we wish to cover a very wide wavelength range, it is still necessary to use two. The most common combination now in use is the 1P28 photomultiplier for the ultraviolet and visible, and a PbS cell for the visible, and near infrared. The necessity of changing detectors in the middle of the range is a limitation which instrument manufacturers have to face in much the same way as the source limitation. As a result, it is common practice for designers to provide either an interchangeable detector mounting plate, a sliding positioner, or a pivoted mirror to direct the beam to the appropriate sensing element.

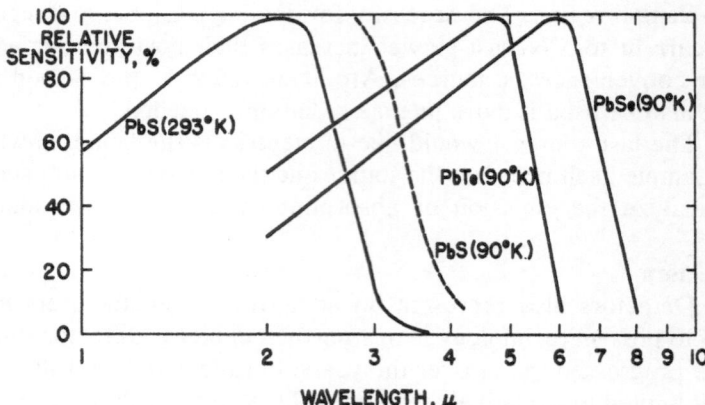

Fig. 4. PbS response curves.

Monochromators

The next section of the instrument on which a decision must be made is the monochromator. The primary questions here are what systems and optical materials to use. Reflective systems using gratings and prisms, or both, are in common use, and each of these has peculiar advantages. In general, suitable gratings give higher dispersion than prisms over a broader wavelength range. However, it is not sufficient to use only a grating to obtain a high-dispersion spectrum. Gratings have the inherent problem of stray radiation from overlapping orders. This can be combatted by using either filters or an order sorting prism.

Prisms, while they do not have stray light problems to the same degree as do gratings, do have the property of nonlinear dispersion. The spectral band width of the emergent beam will therefore vary with wavelength unless the slit opening is also varied with wavelength. In such a monochromator, steps must be taken to vary the slit width to produce essentially a constant spectral band width, whereas in a pure grating monochromator a fixed slit or constant band pass may be used when working in a single order.

For work in the ultraviolet, and especially the far ultraviolet, prisms provide excellent dispersion, and good resolution may be achieved in the ultraviolet and far ultraviolet using a single quartz prism. It is possible to achieve resolution of the order of 0.05 mμ at 180 mμ with this type of system [5].

By combining the two in a foreprism – grating double monochro-

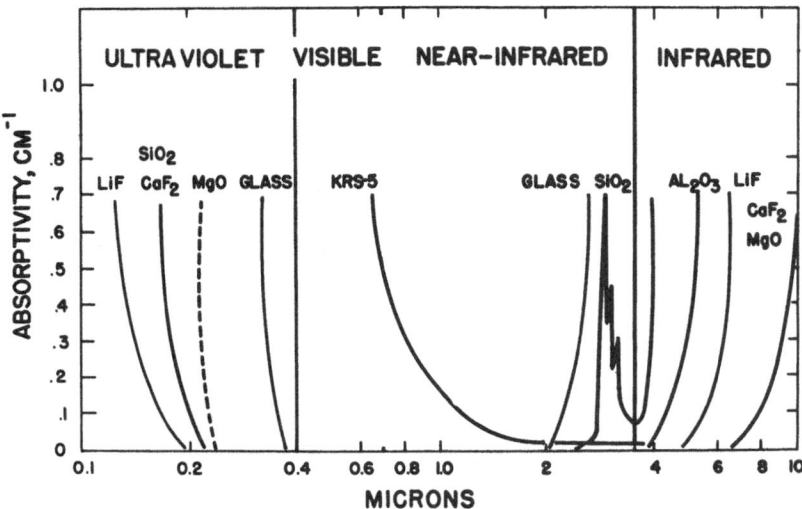

Fig. 5. Absorptivity of various optical materials.

mator arrangement, one may capitalize on the desirable features of both. In the ultraviolet and far ultraviolet, the prism provides high dispersion; in the visible and near infrared, the grating provides most of the dispersion; and throughout the whole region, the prism acts as an order sorter to eliminate stray radiation.

Numerous materials have been used as prism materials. The major factors which must be considered in selecting a material as a dispersing element are absorptivity and dispersion. Obviously it is most desirable to use a prism with very high dispersive power which is also highly transparent. It is usually necessary to draw a compromise here, since the ideal combination is not realized in commonly available materials. Absorptivity and dispersion curves for a number of typical optical materials are shown in Figs. 5 and 6. [6].

The most common and best general-purpose prism material for the ultraviolet – visible – near-infrared regions is quartz. The only real limitation it poses is its own purity. Figure 7 illustrates the comparison between natural and synthetic quartz with respect to their transmittance characteristics. Natural quartz is good down to about 200 – 205 mμ. Beyond this, impurities and inhomogeneities in the quartz absorb and scatter radiation. With careful selection, natural quartz may be used to 185 mμ, but if the object is to work below 185 mμ, either fused silica or synthetic quartz must be used. Both

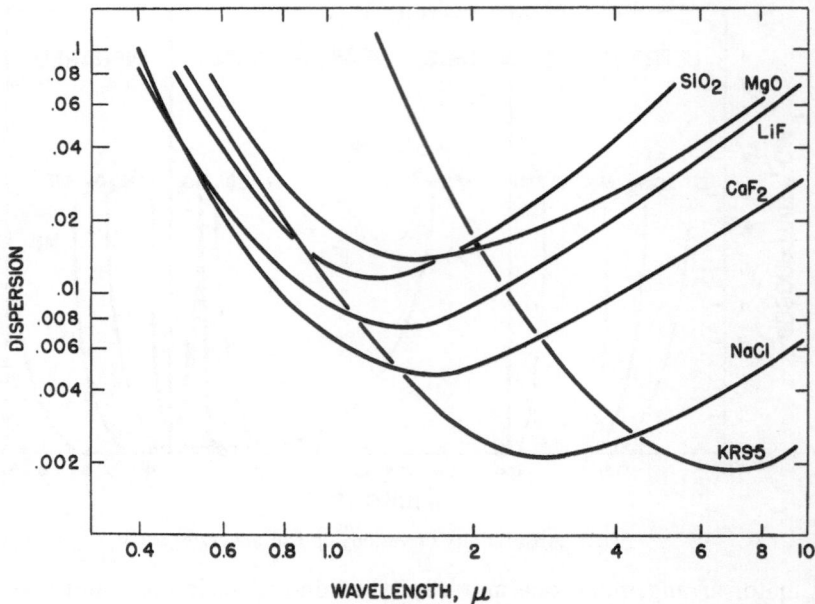

Fig. 6. Dispersion of various optical materials.

of these materials are excellent in the far ultraviolet, synthetic quartz being usable below 160 mμ. The principal limitation here is in the near-infrared region. Fused silica is limited to about 2650 mμ because of entrained water which gives rise to strong hydroxyl absorption bands in the near infrared. Synthetic quartz does not have this trapped water and consequently may be used as far as 3500 mμ.

If the instrument is to cover only the visible region, borosilicate glass gives acceptable transmission from about 325 mμ through the visible region and partway into the near infrared. Figure 8 illustrates the optical diagram of a glass prism spectrophotometer [7] and serves to show how a particular technological development may affect the design of an instrument. The prism in this instrument is a Fery prism, which, unlike a flat prism, acts as its own focusing lens. However, since the focal length of a lens changes with wavelength, a special mount is needed to move the prism laterally as well as turn it on its axis so that it remains in proper focus at the exit slit.

Whether prisms or gratings are chosen as the optical material, there is still another fundamental problem to cope with, the wavelength-dependent variations in source output, detector sensitivity,

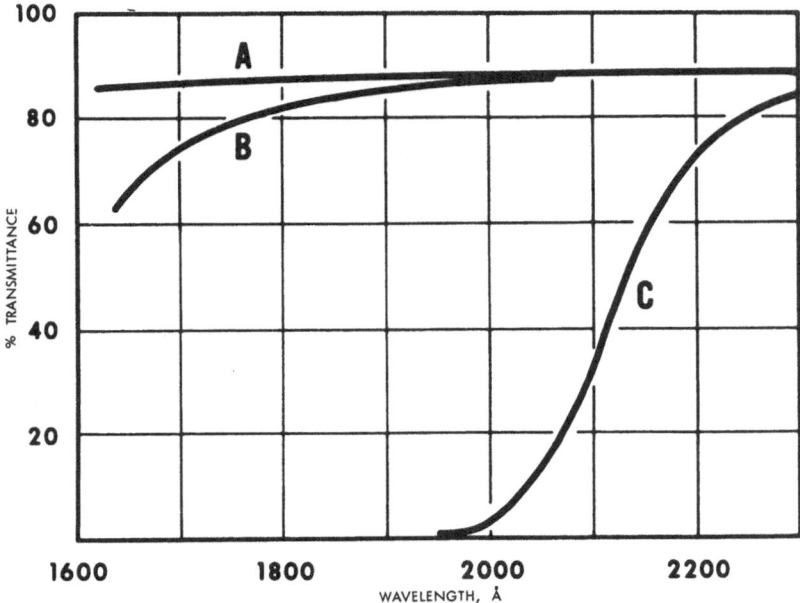

Fig. 7. Transmittance characteristics of quartz. (A) Theoretical reflection loss; (B) synthetic quartz; (C) natural quartz. Path 1.0 in.

prism dispersion (or grating efficiency), and background absorption. These combine to produce a system in which the overall energy level varies with wavelength.

If a reasonably constant response is to be maintained, it is necessary to compensate for this variation. This may be done in one of two ways—optically or electronically. In a manually operated unit such as the DU Spectrophotometer, it may be done optically by setting the wavelength and sensitivity control and adjusting the slit opening to give a 100% T reading. Or, we can set the wavelength and slit opening and vary the sensitivity setting to give a 100% T reading. There are compromises to be made in both these modes of operation. In the former, the compromise is on resolution, to minimize noise level. In the latter, the compromise is on noise level, to gain resolution or sensitivity. The mode of operation chosen depends on the requirements of the particular analysis being made.

In an automatic recording instrument, optical compensation may be accomplished by varying the slit width as a function of

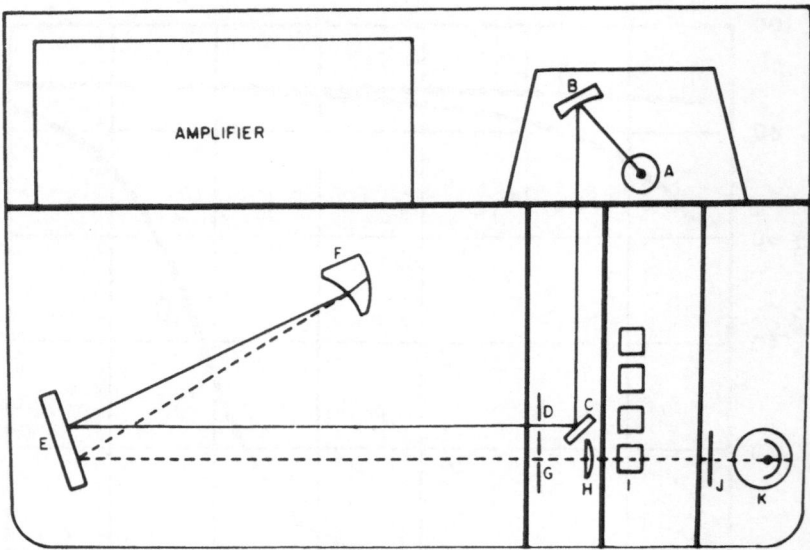

Fig. 8. Model B optical diagram.

wavelength or background energy, using either a cam-programmed
slit system or a slit servo system. Electrical compensation may be
accomplished by varying the amplifier gain and time constant or
by varying the scanning speed as a function of background energy.
In a ratio-recording quartz prism spectrophotometer reference
beam energy is maintained constant by taking the signal from the
reference beam and feeding it through a reference signal amplifier
to a servomechanism which continually adjusts the slit width
to maintain a constant reference energy level. In double-beam
operation, as source intensity or detector response varies, the slit
follows the variation in reference energy.

For example, in differential analysis where the reference material absorbs strongly, or while operating in the water region of
the near infrared, the slit servo system acts to provide a constant
reference energy.

In a pure grating system using linear dispersion or constant
band-pass, the slit opening is fixed. System energy fluctuations
caused by source and detector nonlinearity and background absorption are not compensated manually and must therefore be compensated electrically by automatically controlling amplifier gain,
sensitivity of detector, or scanning speed.

The Sample Compartment

The sample compartment is the point at which the decision is made on whether to build a single-beam or a double-beam instrument, and in the case of a double-beam system, it is usually the place where beam switching or light modulation is performed. It is also the point at which accessory versatility is usually incorporated. Here provisions may be made for special cell arrangments such as temperature-regulated cells, flow cells, reflectance attachments, or special external optics. The physical layout of the instrument to a great extent determines how much versatility can be built in, and the prior choice of one has much to do with fixing the other.

Readout Mechanism

The readout section of the instrument is the one where the greatest amount of variation exists among manufacturers. Here the choice is made among null electrical balance, ratioing, direct reading, and optical null. Whether or not it is to be a recording instrument, and if it records, whether it uses a strip chart, a flat bed, or a drum recorder must also be decided. Here also the choice of speed of operation, flexibility of readout, and fundamental accuracy of the readout system is made.

On the basis of existing instruments, one can generalize that high performance and high speed are incompatible. To use the DU Spectrophotometer as an example, it is a single-beam null-balance instrument possessing very high photometric accuracy. Readings are taken point by point, and at each point the operator must make several manual adjustments. Important here is the fact that the readings are static readings and are inherently more accurate than dynamic readings on the same instrument, since the manual instrument is at statistical equilibrium when the reading is taken. In addition, the null electrical balance method is perhaps the most accurate means of making a transmittance measurement because accuracy depends principally on slidewire linearity, and linearities of better than 0.1% are quite readily accomplished. The necessity of manually adjusting the instrument settings, however, places a limitation on the speed with which measurement can be made.

In certain routine applications, it may be possible to sacrifice some of the extreme accuracy of a manual null-balance instrument for the increased speed of a manual direct-reading instrument with a meter readout. Incorporation of an adapter to eliminate the need

Fig. 9. Model B test tube adapter.

for transferring solutions from test tube to absorption cell is an aid in gaining rapidity of operation. An example of this is shown in Fig. 9. The limitation which this places on the manufacturer is twofold. The first is meter accuracy. Meters are limited to accuracies of about ±1%. In applications where photometric accuracy requirements of this order are permissible, however, meter readout is often adequate. The second limitation is amplifier stability, or the ability to make successive readings over an extended period without frequent adjustment of the controls. Feedback-stabilized amplifiers provide this stability and are commonly used.

The next obvious step in the "performance *vs*. convenience" scale is a recording instrument with the same performance characteristics as the nonrecording instrument on which it is based. The comparison chosen here is between the DU and DK Spectrophotometers. The increased complexity of the DK is a result of a decision to optimize the system to obtain speed and convenience and still maintain the necessary degree of accuracy for analytical measurements.

If all the performance characteristics of a nonrecording instrument could be transferred to a recording instrument without loss, it would indeed be an ideal situation. In practice, it is often difficult to buy back speed and convenience without loss in some area of performance. Necessary changes in the layout of the various components to get the degree of convenience and versatility desired can cause significant differences in the results obtained on the two instruments. This is one of the major reasons why discrepancies exist in measurements on the same material made with instruments of different design. These factors do not only manifest themselves among instruments made by different manufacturers, but will also appear when comparing measurements on the same sample using instruments of different design made by the same manufacturer.

APPLICATIONS OF NEAR-INFRARED SPECTROSCOPY

Qualitative Analysis

The near infrared has not been exploited for qualitative analysis to the same extent as the fundamental region, but much valuable work has been done along these lines. In his classic review published in 1953, Kaye [3] presented a Colthup-type group frequency chart for, CH, OH, and NH overtones and combinations between 0.6 and 3.4 μ based on the spectra of over 100 purified compounds plus the compilation of literature values. A portion of this chart is shown in Fig. 10. A later publication by Goddu [8] is somewhat more extensive and is perhaps the most comprehensive work to date on characteristic group frequencies in the near infrared.

Molecular Association Studies

Hydrogen Bonding. The phase of near-infrared spectroscopy which to date has received the most attention has been the study of inter- and intramolecular association phenomena.

The most familiar type of association phenomenon is referred to as "hydrogen bonding" and is caused by asymmetrical charge

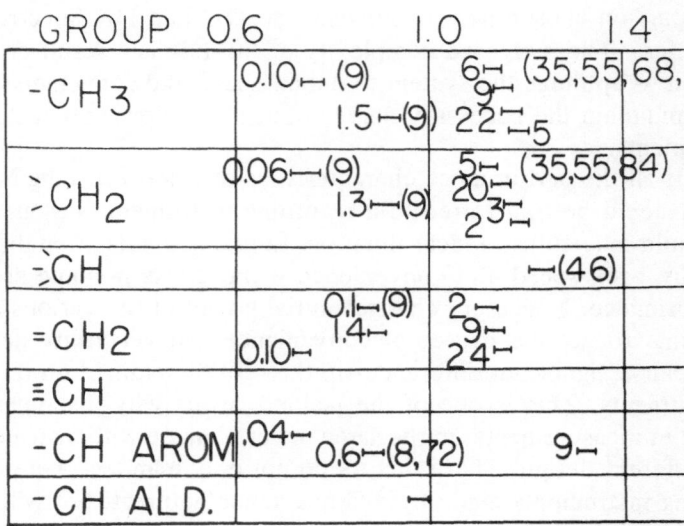

Fig. 10. Near-infrared group frequency chart.

distributions within molecules or functional groups, which produce electrostatic effects either between neighboring molecules or between neighboring functional groups in the same molecule.

These effects generally cause perturbations in the position and intensity of bands characteristic of the groups involved in the interaction.

The association of neighboring molecules is referred to as intermolecular hydrogen bonding and typically is observed in highly polar molecules such as alcohols, phenols, and acids. It is characterized by broad intense bands which shift to shorter wavelength and become much sharper upon dilution with a nonassociating solvent. This effect is illustrated in Fig. 11. The alcohol concentrations and path lengths were selected to maintain the same number of solute molecules in the beam at all times. The increase in intensity of the free OH (2.75 μ) and its first overtone (1.41 μ) is quite pronounced on dilution. This illustration makes it apparent that if one wishes to study the associated (bonded) state, the fundamental region is the best. If the main interest is in the free (unbonded) state, the overtone region should be used.

A second example of intermolecular association is illustrated in Fig. 12. The interaction of Lewis bases and *n*-amyl alcohol in

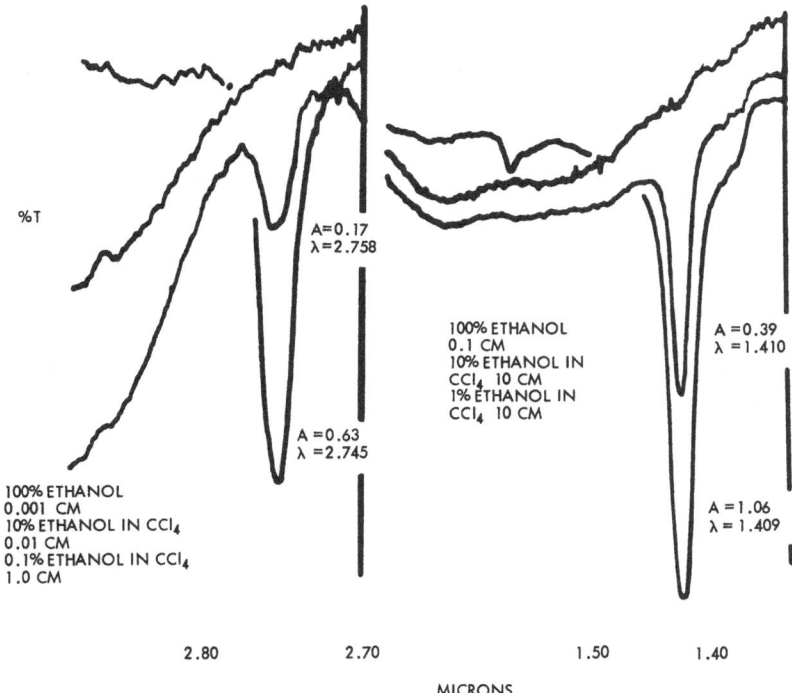

%T

A=0.17
λ=2.758

A =0.63
λ =2.745

100% ETHANOL
0.001 CM
10% ETHANOL IN CCl₄
0.01 CM
0.1% ETHANOL IN CCl₄
1.0 CM

100% ETHANOL
0.1 CM
10% ETHANOL IN
CCl₄ 10 CM
1% ETHANOL IN
CCl₄ 10 CM

A =0.39
λ = 1.410

A =1.06
λ = 1.409

2.80   2.70   1.50   1.40

MICRONS

Fig. 11. The effect of concentration on hydrogen bonding in ethanol.

CCl_4 solution causes a shift in wavelength of the complex band, which appears to bear no relationship to the concentration of alcohol – base complex.

The second type of association, that between adjacent functional groups in the same molecule, is referred to as intramolecular association. A typical example of this effect is illustrated in the infrared spectrum of o-nitrophenol. The phenolic OH band is perturbed in position and intensity by the adjacent nitro group, but there is no change on dilution.

Solvent Effects. A form of molecular association phenomenon which is often observed is a shift in frequency on dilution, caused by the dielectric constant or refractive index of the solvent. This shift $\Delta\nu/\nu$ is governed by the relationship

$$\frac{\Delta\nu}{\nu} = C\,\frac{D-1}{2D+1} = C\,\frac{n^2-1}{2n^2+1} \tag{3}$$

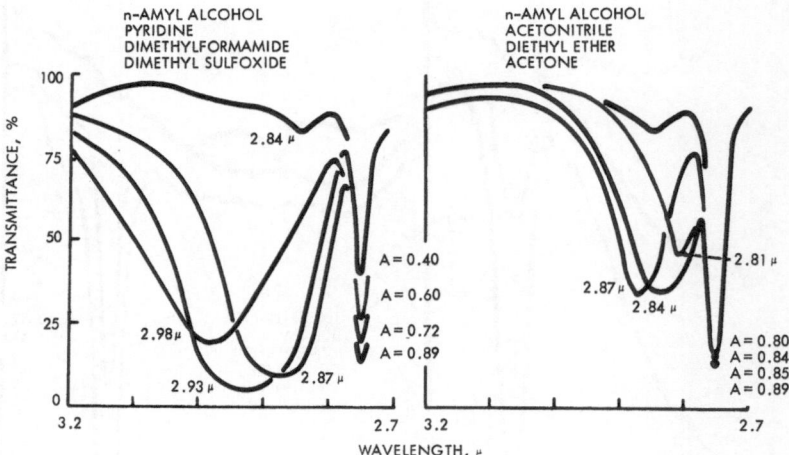

Fig. 12. Infrared spectra of mixtures of *n*-amyl alcohol (0.020 *M*) and Lewis bases (0.20 *M*) in CCl$_4$. (Cell 1.00 cm).

where C is a proportionality constant equal to about 0.06 [9], D is the dielectric constant, and n is the refractive index.

Other solvent effects which might be considered are the perturbations in peak intensity upon changing solvents. Table I shows the variation in peak intensity of the 1.40 μ *n*-amyl alcohol band in various solvents. This illustration shows a large change in peak intensity but a rather minor change in the integrated area associated with changing solvents.

TABLE I
Intensity of 1.40-μ *n*-Amyl Alcohol Band in Various Solvents

Solvent	ϵ_x	ϵ_x Breadth
Fluorolube	1.60	0.024
n-Hexane	1.90	0.024
Carbon tetrachloride	1.75	0.020
Carbon disulfide	1.47	0.019
o-Dichlorobenzene	1.36	0.019
Chlorobenzene	1.22	0.019
Bromobenzene	0.12	0.019

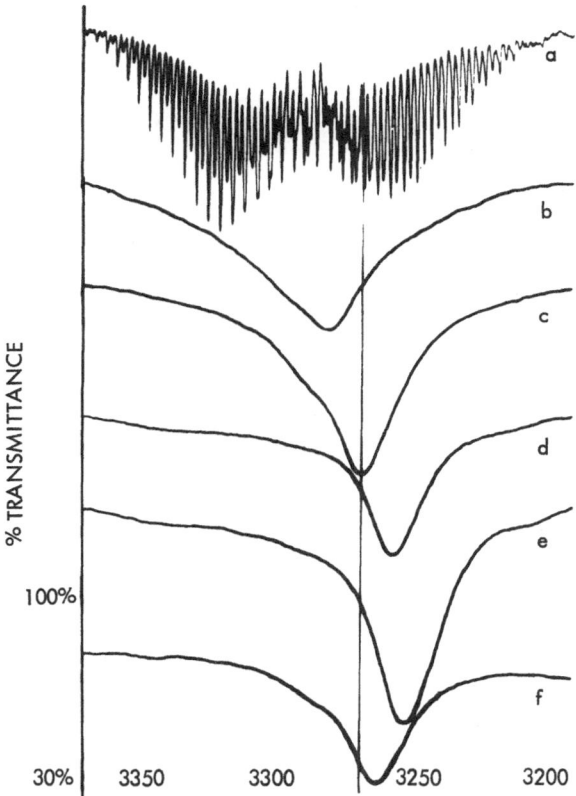

Fig. 13. The effect of solvation on Fermi resonance of C_2H_2 fundamental. (a) Acetylene cell length 100 mm, pressure 25 mm, ν_{max} = 3282.5; (b) in Fluorocarbon Oil No. 180, $\Delta\nu$ = 3.0; (c) in carbon tetrachloride, $\Delta\nu$ = 16.0; (d) in chlorobenzene, $\Delta\nu$ = 27.5; (e) in iodobenzene, $\Delta\nu$ = 31.2; (f) in 2 parts carbon tetrachloride and 1 part iodobenzene, $\Delta\nu$ = 22.5.

Solvation of a gas in a solvent can also cause severe perturbation of band positions. This is illustrated by Figs. 13 and 14 which are the fundamental and overtone bands of acetylene as a gas and in various solvents.

The fundamental spectrum, which was run on a Beckman IR-7, shows a good example of Fermi resonance, which, as I mentioned earlier, is quite often observed in the near infrared. The resonance may also be seen on the high frequency side of the band in the fluoro-

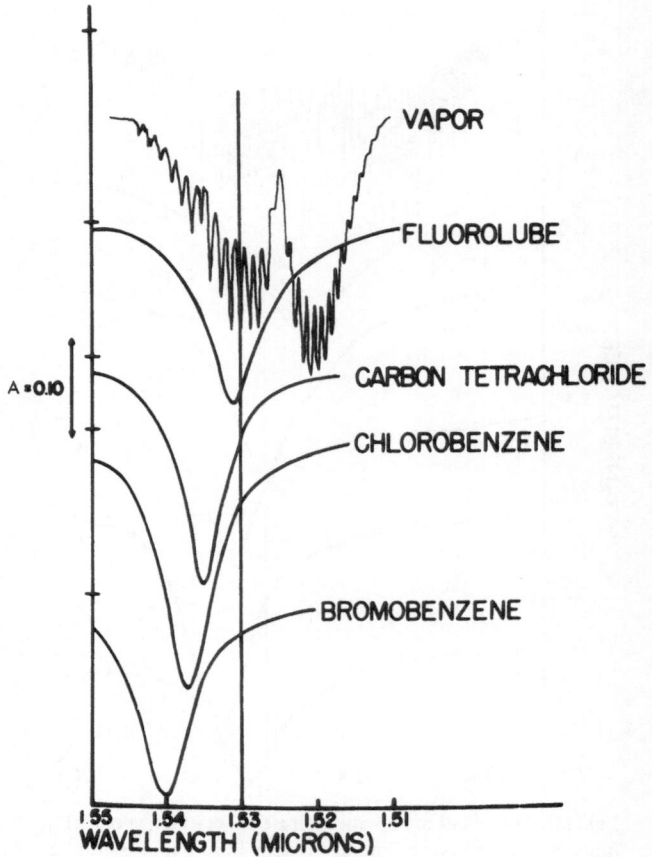

Fig. 14. The effect of solvation of C_2H_2 overtones.

carbon and CCl_4 solution spectra, but is not apparent in iodobenzene, presumably because the iodobenzene shifts the bands enough to move them out of a resonant condition.

Study of Individual Molecules

A number of individual molecules have been studied with emphasis on the assignment of overtone and combination bands. Kaye [3] has published complete assignments of all possible bands to and including some quintary combinations for bromoform, cloroform, methylene chloride, methylene bromide, and benzene. Por-

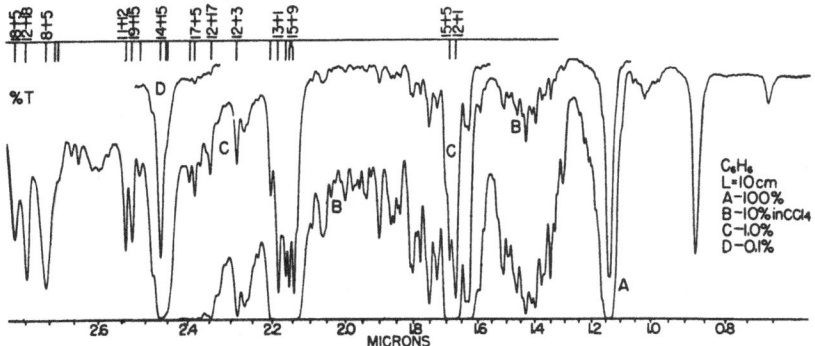

Fig. 15. Near-infrared spectrum of benzene.

tions of the spectra of benzene and bromoform are shown in Figs. 15 and 16 (with band notation as originally published.)

Other molecules which have been studied include fluoroform [9, 10], 1, 1, 1 – trifluoroethane [11], and methyl iodide [12].

Quantitative Analysis

To the practical analyst, the class of near-infrared applications of greatest significance is quantitative analysis. As is also true in most other areas of application, the near-infrared region has not been used in quantitative analysis nearly as much as the infrared. This has been due in the past largely to limitations in available instrumentation. In the last 8 – 10 years, however, commercial instruments capable of producing reliable results have become available, and there has been a resultant increase in the amount of attention paid to near-infrared analysis as a quantitative tool. Re-

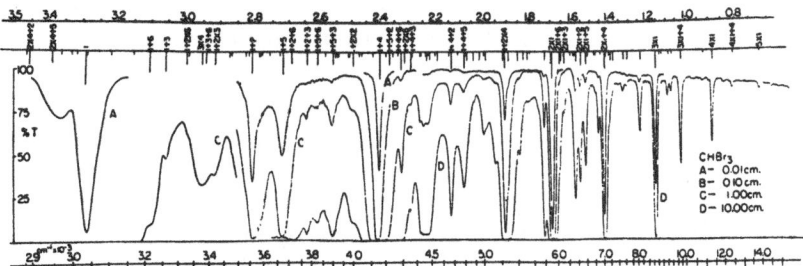

Fig. 16. Near-infrared spectrum of bromoform.

cently, a number of good papers have been published on quantitative analysis of various mixtures [13-21].

The problems associated with applying the near infrared to quantitative analysis are no different nor more difficult than those associated with quantitative analysis in any other region. A knowledge of the spectrum of each component in the sample is always desirable in quantitative analysis of specific mixtures. However, in complex samples, this may not be practical, and prediction of the feasibility of an analysis may be based on bands characteristic of the analyte.

The two main requirements which must be met in setting up an analysis are: (i) a band characteristic only of the analyte must be found and (ii) the absorptivity of this band must be large compared with the sum of absorptivities of the other components of the sample. The first requirement is mandatory; the second may be circumvented if the analyte can be selectively removed from the sample. If it is possible to extract the analyte selectively into a noninterfering solvent, the pure extraction solvent may be used as the reference material. An example of this is the determination of water in gasoline, where ethylene glycol is used to extract the water, and the pure ethylene glycol is used as the reference. If the analyte can be removed only with an interfering solvent, then the base material may be used as a reference after the analyte has been removed.

Sample-Handling Techniques

Sample-handling techniques used in the near infrared are similar in some respects to those used in conventional infrared spectroscopy and in some respects are quite different.

In the region from $3-40\ \mu$, cell thicknesses of from 0.025 to 1 mm are commonly used, since most of the absorption bands observed in this region are fundamental vibrations and are quite intense in comparison with the overtone and combination modes observed in the near infrared.

Selection of the proper cell thickness for near-infrared work depends to some extent on the spectral region of interest. Generally, as the wavelength of interest decreases, the optimum cell path increases, because of the characteristic decrease in intensity of higher-order overtones. Approximate optimum cell thicknesses for various spectral regions are indicated in Table II. (The transmittances of various cell materials are shown in Fig. 5.)

TABLE II
Optimum Cell Thickness

Wavelength, μ	Cell length, cm	Designation
3.5 – 2.7	0.005 – 0.2	Fundamental
2.7 – 1.8	0.05 – 2.0	Combination
1.8 – 1.4	0.1 – 1.0	1st Overtone
1.4 – 0.9	1.0 – 20	2nd Overtone
0.9 – 0.6	10 –	3rd and 4th Overtones

Samples may be examined as gases, liquids (solutions), or solids, either as KBr pellets or Nujol mulls. In work with KBr and Nujol, however, it is necessary to reduce particle size far below the point required for conventional infrared work, and therefore these techniques have not been used as extensively as liquid or solution techniques.

Carbon tetrachloride is nearly ideal as a near-infrared solvent, since it is transparent at paths up to 50 cm. This makes possible the use of dilute solutions, which are usually preferred, especially in the case of molecules exhibiting intermolecular association.

Many of the common infrared solvents may be used as near-infrared solvents over limited regions of the spectrum, and the selection of a solvent for a particular application is based on the same general principles: solvent power for the sample and transmission characteristics in the necessary path lengths and spectral regions. In certain cases, solvents not normally used in the infrared due to their intense absorption may be used to advantage for near-infrared determinations. For example, in the determination of water in jet fuels, ethylene glycol may be used to extract the water. Differential analysis using pure ethylene glycol as reference is used to determine the water content.

Sensitivity of Near-Infrared Determinations

In order to predict the point at which one observes maximum change in transmittance (dT) for a given change in concentration (dc), one must manipulate Beer's law.

The basic expression for Beer's law is

$$T = I/I_0 = e^{-abc} \tag{4}$$

where T is the transmittance, I_0 the incident radiation, I the transmitted radiation, a the absorptivity, b the cell length, and c the concentration. Differentiating (4) with respect to c, one gets

$$\frac{dT}{dc} = -ab\, e^{-abc} \tag{5}$$

at fixed values of b. To find the point of maximum dT, differentiate (5) with respect to b and set the result equal to zero:

$$\frac{d}{db}\frac{(dT)}{(dc)} = -a\left[-abc\, e^{-abc} + e^{-abc}\right] = 0$$

$$a^2bc\, \mathrm{e}^{-abc} + ae^{-abc} = 0$$

Division by $-ae^{-abc}$ yields $-abc + 1 = 0$. Thus, $abc = 1$ and

$$T = e^{-abc} = e^{-1} = 1/e = 36.8\%$$

and b is the cell length at which $T = 1/e$. Using this as a guide, Table III may be used to estimate the sensitivity of some typical determinations. The values in Table III assume a 1% change in absorbance when the cell length is selected to give a transmittance of $1/e$, and when instrument noise level is 1%.

These values do not represent the ultimate in sensitivity attainable with near-infrared techniques. Incorporation of concentration steps or selective extraction of a large amount of sample with a small amount of solvent can, in favorable cases, lower some of these values by as much as a hundred times.

SUMMARY

This brief discussion outlines some of the fundamentals of near-infrared spectroscopy. Bands representing overtones of CH, NH, and OH stretching vibrations are observed in this region and are useful in theoretical, qualitative, and quantitative studies on a wide variety of materials.

Improved instrumentation and continuing study of the capabilities of the near-infrared as a spectral region should provide further expansion of its utility as an analytical tool.

TABLE III
Typical Sensitivity of Near-Infrared Determinations

System	λ, μ	Sensitivity, %
Water in hydrocarbons	2.70	0.004 – 0.02
	1.9	0.02 – 0.06
	1.4	0.2 – 0.6
Water in alcohols	2.70	0.05 – 0.2
	1.9	0.035 – 0.08
Water in carboxylic acids	1.4	0.04
Alcohols in hydrocarbons	2.76	0.03 – 0.10
(diluted in CCl_4)	1.4	0.1 – 0.3
Alcohols in acids	2.76	0.2 – 0.5
(diluted in CCl_4)	1.4	0.5 – 1.0

BIBLIOGRAPHY

For an extensive bibliography on all aspects of near-infrared spectroscopy, the reader is referred to references [4] and [7] and to *The Encyclopedia of Spectroscopy*, edited by Clark, and published by Reinhold.

Instrumentation

Cahn, L., and Henderson, B. D., "Performance of the Beckman DK Spectrophotometer," *J. Opt. Soc. Am.* 48:380 (1958).

Phillips, J. P., "Use of the Beckman DU Spectrophotometer in the 1.0 – 1.9 Micron Range," *Anal. Chem.* 31:1604 (1959).

Shuler, W. E., "Near Infrared Spectroscopy with the Beckman DU Spectrophotometer," *Anal. Chem.* 31:1604 (1959).

General

Andrillat, Y., "Spectrophotometric Study of Wolf-Rayet Stars in the Near Infrared," *Compt. rend.* 237:784 (1953).

Barchewitz, P., "Use of Near Infrared in Chemical Analysis," *J. Chem. Phys.* 45:40 (1943).

Barr, E. S., and Harp, W. R., "The Near Infrared Absorption Spectra of Some Vegetable Oils," *Phys. Rev.* 63:457 (1943).

Freymann, M. and R., "On the Identification of Homologous Isomeric Organic Compounds by Their Near Infrared Absorption Spectra," *Compt. rend.* 213:174 (1944).

Freymann, M. and R., "Identification of Fuels by Near Infrared Spectroscopy." "Application to the Analysis of Ketones," *Groupe. franç. dévelop. recherches aéronaut. Note. technique* 11:53 (1943).

Freymann, R., "On the Question of Intra- and Inter-molecular 'Hydrogen Bonding' and Its Absorption Spectra in the Near Infrared," *J. Chem. Phys.* 6:497 (1938).

Holman, R. T., and Edmondson, P. R., "Near Infrared Spectra of Fatty Acids and Related Substances," *Anal. Chem.* 28:1533 (1956).

Howard, J. N., and Chapman, R. M., "The Pressure Dependence of the Absorption by Entire Bands of Water Vapor in the Near Infrared," *J. Opt. Soc. Am.* 42:423 (1952).

Lauer, J. L., and Rosenbaum, E. J., "Near Infrared Absorption Spectrophotometry," *Appl. Spectry.* 6:29 (1952).

Mizushima, S., *et al.*, "The Molecular Structure of N-methyl Acetamide," *J. Am. Chem. Soc.* 72:3490 (1950).

Mizushima, S., *et al.*, "Near Infrared Spectra of Compounds with Two Peptide Bonds and the Configuration of a Polypeptide Chain. V.," *J. Am. Chem. Soc.* 76:6003 (1954).

Sanders, C. L., and Middleton, E. E. K., "Absolute Spectral Diffuse Reflectance of MgO in the Near Infrared," *J. Opt. Soc. Am.* 43:58 (1953).

Studer, F. J., and Cusano, D. A., "TiO_2 Films as Selective Reflectors in the Near Infrared," *J. Opt. Soc. Am.* 43:522 (1953).

White, L., Jr., and Bassett, L. G., "Determination of Water in Burning Nitric Acid by Near Infrared Absorption," *Anal. Chem.* 28:1538 (1958).

Whetsel, K., Roberson, W. E., and Krell, M. W., "Near Infrared Spectra of Primary Aromatic Amines," *Anal. Chem.* 30:1598 (1958).

REFERENCES

1. Herzberg, G., Infrared and Raman Spectra, D. Van Nostrand Co., New York, 1945.
2. Gauthier, G., *J. phys. rad.* 14:19 (1953).
3. Kaye, W. I., *Spectrochim. Acta* 6:257 (1954).
4. McCallum, J. D., *Trans. N. Y. Acad. Sci.* 24:140 (1961).
5. Beckman Instruments, Inc., Bulletin 735-A.
6. Kaye, W. I., *Spectrochim. Acta* 7:181 (1955).
7. Miller, et al., *J. Opt. Soc. Am.* 39:377 (1949).
8. Goddu, R. F., *Anal. Chem.* 32:140 (1960).
9. Jones, L. H., and Badger, R. M., *J. Am. Chem. Soc.* 73:3132 (1951).
10. Bernstein, H. J., and Herzberg, G., *J. Chem. Phys.* 16:30 (1948).
11. Rix, H. D., *Ibid.* 21:1077 (1953).
12. Coggeshall, N., *Anal. Chem.* 22:381 (1950).
13. Herzberg, G., and Herzberg L., *Can. J. Research* 278:322 (1949).
14. Goddu, R. F., *Anal. Chem.* 30:2009 (1958).
15. Fernandez, J. E., *et al.*, *Ibid.* 32:158 (1960).
16. Goddu, R. F., and Delker, D. A., *Ibid.* 30:2013 (1958).
17. Brandenberger, H., and Bader, H., *Ibid.* 33:1947 (1961).
18. Whetsel, K., Robertson, W. E., and Krell, M. W., *Ibid.* 30:1594 (1958).
19. Powers, R. M., Harper, J. L., and Tai, H., *Ibid.* 32:1287 (1960).
20. Crisler, R. O., and Burrill, A. M., *Ibid.* 31:2055 (1959).
21. Kyriacou, D., *Ibid.* 33:153 (1961).
22. Cordes, H. F., and Tait, C. W., *Ibid.* 29:485 (1957).

Progress in Ultraviolet and Visible Spectrometry

Robert G. White

Tests and Inspection Dept.
National Aniline Division
Allied Chemical Corporation
Buffalo, New York

---

1. INTRODUCTION

This review is intended to update by approximately one year that which appeared in Vol. 1 of *Progress in Infrared Spectroscopy.* It is confined to electronic absorption spectra and does not include the near-infrared region.

1.1 Activity in this field continued at a high level. In my literature search I encountered many interesting applications, but few that rate as startling.

1.2 As Hirt pointed out in his excellent review [58], there is a trend toward greater sophistication in ultraviolet spectrometry. This includes an increasing dependence on differential spectra.

2. HISTORICAL

Discussions of the laws of absorptiometry continue fashionable. Swinehart's paper on the Beer-Lambert law [138] is worth reading. Malinin and Yoe [83] presented a definitive survey of the laws of Bouguer and Beer, and their relationship to the work of Lambert and Bernard. Morton [102] reviewed the development of ultraviolet and visible spectrometry.

3. TECHNIQUES

3.1 *Spectrum presentation*: Tunnicliff and Hawes [143] described an absorptivity recorder for the Cary spectrophotometer.

3.2 *Solvents*: Johnson [64] discussed the purification of hydrocarbon solvents. A "solvent" good to 220 mμ was proposed by McDonald and Cook [90] – stir the sample into melted polyethylene and press into a film. Brierley and Langbridge [20] told how to prepare spectro-grade alcohol.

3.3 *Absorption cells*: Breda and Kotkas [19] described a simple variable-thickness cell.

3.4 *Optimum absorbance*: Wybourne [159] considered this for the case of shot-noise-limited instruments.

3.5 *Instruments*: Grum and Scharf [48] extended a G.E. spectrophotometer into the ultraviolet. A device for monitoring a chromatographic column at 276 mμ to determine *o*-, *m*-, and *p*-cresol was described [2]. Hirt *et al.* [59] discussed the spectral energy distribution of Beckman, Nester, and other ultraviolet sources.

   3.5.1 A discussion of the present status of commercially available hardware might well be obsolete before printed. I expect big changes soon, due to the development of better sources (e.g., a deuterium discharge lamp), detectors (the end-on multiplier phototube), and optics. There is a big need for a derivative spectrophotometer, but the demand has not yet been great enough to tempt the instrument manufacturers.

3.6 *Vapor-phase work*: Kaye [67, 68] discussed the far-ultraviolet spectrophotometer as a gas chromatograph detector. Vapor-phase spectra are popular in the far ultraviolet because of the high absorptivities of analytes and the low transmittance of solvents.

McGrath, *et al.* [91], published a long, comprehensive review of the analysis of gases and vapors by spectroscopic techniques. They included some theory, data on benzene, halobenzenes, alkylbenzenes, amines, phenols, and heterocyclics; also given are application data and the limits of detection of certain organic solvents.

3.7 *Reflectance*: McAloren [89a] described means of preparing reproducible MgO standards by electrostatic deposition of MgO smoke on silver disks. He pointed out that MgO disks prepared by compacting at 1000 psi show decreased reflectance in the ultraviolet and develop absorption bands in the near-infrared.

The photostability of tablets colored with FD&C Yellow No. 5, Blue No. 1, and Red No. 3 was evaluated by reflectance [75].

3.8 *Solid state spectra*: Ultraviolet spectrometrists are at last discovering the pressed-disk technique. KCl disks of 1 mm thickness have been found to have more than 60% transmittance from 320 to 220 mμ [127]. Waggoner and Chambers also discussed the ultraviolet spectrometry of solids [151].

3.9 *Ultraviolet spectrometry by polarized light*: Dörr and Held reviewed this [35]; they give 88 references.

3.10 *Photochromism*: Wettermark established the photochromism of o-nitrotoluene, 2,4-dinitrotoluene, and certain of their derivatives [154], using flash photolysis at −70°C. He explains it by:

Colorless   Colored

3.11 *Electrochromism*: Platt [115] showed that the absorption and emission spectra of certain dyes shifted several hundred angstroms upon the application of a strong electric field. Labhart [74] conducted similar studies.

3.12 *Differential spectrometry*: This is now so routine, particularly among pharmaceutical analysts, as to scarcely merit special mention. Ross and Wilson [120] described the determination of a single component. It has also been applied in the determination of niobium [2a], titanium [84], neodymium [85], phenobarbital [142], thalidomide [46], and hexachlorophane [37].

3.13 *High-temperature spectrometry*: Carnall [27] studied the visible and near-infrared spectra of the lanthanides in molten $LiNO_3 - KNO_3$ eutectic at 150°C.

3.14 *Kinetic studies*: Holder [61] showed the utility of spectrometric techniques in a kinetic study. Wood and Gilford [158] recorded absorbance *vs.* time at fixed wavelength to study enzyme-catalyzed reactions. They presented velocity recordings of both these and some chemical reactions.

3.15 *Calculations*: Sternberg's technique was applied by Vasilenko *et al.* [149a] in a nine-wavelength multicomponent method. Niebergall and Mattocks discussed [109] multiple regression in multicomponent analysis. The theory underlying such analyses was discussed by Grinev *et al.* [47]. Pillion *et al.* [114] treated absorptivity corrections in multicomponent analysis — both corrections for nonadditivity and deviations from Beer's law. They gave as examples a ternary mixture of phenols in the infrared region and visible spectra of a nonadditive mixture of two dyes. Garrett *et al.* [42a] used an analog computer to compare the effects of substituents on the ultraviolet spectra of 4-ene-3-one steroids and to provide a spectrum library useful for characterizing substituents. Pernarowski *et al.* [111] discussed the theory of analysis of binary mixtures via absorbance ratios. The ratio is a nonlinear function of composition, but is independent of dilution. It is a rectilinear function when one analytical wavelength is an isosbestic point. They extended the technique to include ternary mixtures. Blumer [17] contrived a slide rule for spectrophotometric calculations. Stanford [135] devised a figure of merit for the spectrometric analysis of small samples. This is LT/V, where L is the length, V is the volume occupied by the sample, and T is the fraction of the sample-beam energy which is transmitted by the sample-holding system when no analyte is present. LT/V is inversely proportional to the amount of the sample required for an analysis.

3.16 *Far ultraviolet*: The greatest advances in the ultraviolet field are taking place in the Schumann region. Wilbur Kaye's noteworthy pioneering in the near infrared is now being duplicated by him in the far ultraviolet. His recent publications include a review on analytical aspects, with 256 references [69], far-ultraviolet spectra of H-bonded methanol [70], and two papers [67, 68] on the far-ultraviolet spectrophotometer as a gas chromatograph detector. Price reviewed the vacuum ultraviolet (79 references) [117]; Milazzo reviewed analytical possibilities of the Schumann region [93]; and Ulrich [144] described some far-ultraviolet studies of biologically-active substances.

3.17 *Dissociation constants*: An ultraviolet method for the determination of the pK of meconic and chelidonic acids [97] should have quite general applicability. Edward and Wang [36] discussed the determination of the ionization ratio of thioacetamide in aqueous NaOH. Chen and Laidler [28] presented an ultraviolet method for estimating pK and reported data on phenol, cresols, and xylenols at various temperatures from 5 to 38°C. Burkhard [24] measured the ionization constant of dyes.

3.18 *Miscellaneous*: Sill [129] presented visible transmittance spectra of 55 glass filters, 35 gelatin filters, and 98 two-filter combinations. This is a valuable collection of data for those of us who still occasionally use this primitive monochromator (and who does not?)

4. ERRORS IN SPECTROMETRY

4.1 *Philosophy*: Feldmann [39] stated that it had been suggested that the next editions of the U.S. Pharmacopeia and the National Formulary include the sentence: "It is absolutely essential to have a thorough knowledge and understanding of the limitations as well as the capacities of the instrument before operation of a spectrophotometer is attempted." I heartily concur.

4.2 *Reproducibility*: Phillips [113] gleaned the replicates from data from 20,000 spectra (vintage 1958–1959). He found wavelength reproducibility to be good. The range for four out of five maxima was 2 mμ or less. On the other hand, 20% of log ϵ values had a range of 0.10 or more (a log ϵ difference of 0.10 corresponds to an ϵ ratio of 1.26 to 1, Phillips pointed out).

4.3 *Stray light*: Fog [40] proposed the extinction coefficient at zero concentration as an estimable parameter when stray light is a problem or when Beer's law is not followed.

4.4 *Finite slit*: Lempicki [78] discussed errors arising from this cause.

4.5 *Photometric and wavelength inaccuracy*: This was discussed by Tereshin [141], as was the accuracy of differential methods. Vandenbelt [148] described the holmium oxide filter. This has bands suitable for wavelength calibration at 241, 250, 260, 279, 288, 333, 360, 419, 453, 537, and 638 mμ. Certified filters are available from the N.B.S.

5. BOOKS AND REVIEWS

5.1 *Books*: Rao's book [118] on ultraviolet and visible spectroscopy is too brief for the wide scope it attempts. Hampel's [53], in German, offers a general survey of the ultraviolet and visible which is probably worth the effort only if the reader is at home in that language. Lang's [76] is a looseleaf collection of spectra, mostly of Hungarian origin. It includes 25 pages of conversion tables and a 55-page theoretical introduction. Volume 1 (in two parts) has 414 pages of spectra. This is evidently intended as a continuing service and is worthwhile for those who cannot afford not to have complete information. Bauman [11] offers a good, general treatise on the ultraviolet and visible regions. The new book by Jaffé and Orchin [62] is a unified treatment of ultraviolet spectrometry in terms of a qualitative molecular orbital theory and includes a lot of empirical knowledge of ultraviolet spectra as well. Hershenson's second index, of ultraviolet and visible spectra, which covers the period 1955–1959 [57] is a must. Beginners should read the first half of the book by Beaven *et al.* [13], as well as the discussion on spectrophotometry by Brode and Corning [21] in Berl's book.

5.2 *Reviews*: The biennial fundamental reviews in *Analytical Chemistry* on light absorption spectrometry [92] (Mellon and Boltz, period September 1959 to October 1961; 535 references) and ultraviolet spectrometry [58] (Hirt, 166 references), are, as always, excellent. For an elementary review of spectrophotometry (basic principles and instruments), see Sawyer [122]. A more sophisticated review is that of Mason [89] on molecular electronic absorption spectra – 483 references.

6. APPLICATIONS

6.1 *Anions (inorganic)*: Underwood and Howe [145] determined CO_3^{2-} by a spectrophotometric titration at 235 mμ. They showed spectra, 190–270 mμ, of 0.01 N $Na_2 CO_3$, CO_2, and $NaHCO_3$. The latter two are, of course, essentially transparent. Sulfur dioxide has been determined at 208 mμ, using the plumbous ion [16]. Visible spectra of iodine in oleum were published [6]. The ubiquitous barium chloranilate has now been applied [132] in the determination of boron at 355 mμ.

6.2 *Acids, carboxylic*: Hartford [55] determined citric, itaconic, aconitic, and fumaric acids at 385 and 435 mμ after reacting them with pyridine and acetic anhydride. Benzoic acid was determined in phthalic anhydride by a chloroform extraction at pH 4.00, followed by measurement at 274 mμ [105].

6.3 *Alcohols*: Crummett [30] described an improvement on Shmulyakovskii's method [128]. He determined the near-ultraviolet spectra of nitrites of butoxyl alcohols. The ratio $A_{355\ m\mu}/A_{385\ m\mu}$ was found useful in determining the primary to secondary hydroxyl ratio on polypropylene glycols.

6.4 *Amines*: Baba discussed the electronic spectra of aniline [8], and naphthylamines [9]. Slifkin showed ultraviolet and visible spectra of complexes of triethylamine, diethylamine, and ethylamine with chloranil in 50% alcohol [130]. Burkhard [24] reported K_α of p-aminoazobenzene to be 1.26×10^{-3}. Gerson and Heilbronner [42b] discussed the structure and spectra of protonated forms of p-dimethylaminoazobenzene and ten p'- and m'-substituted derivatives, as well as p,p'-bisdimethylaminoazobenzene. They derived regression equations expressing the relationship between bathochromic shifts on protonation and relative basicity. Feldman [38] determined N-phenyl-1-naphthylamine in oils by acid extraction, transfer to cyclohexane, and measurement at 338 mμ.

6.5 *Biochemical applications*: These are so plentiful that those discussed here are but a random sampling designed more to intrigue than to inform. Beaven reviewed the ultraviolet spectrometry of proteins and related compounds [12]. Glazer and Smith discussed ultraviolet differential spectra obtained on denaturation and proteolysis of proteins and polypeptides [43]. Hamilton briefly described electronic spectra as tools in biochemical analysis [52]. Morton's review paper on spectroscopy as a biochemical tool [103] was principally concerned with the ultraviolet region. He included mentions of the determination of molecular weight and the usefulness of isosbestic points. Ultraviolet spectra and reaction kinetics in biochemistry were considered in a review by van Dranen [149]. Turning to more specific papers, the characterization of singly-unsaturated steroids in the 170–210 mμ region was described [23a]; Wilson [156] used absorption spectra in the 200–700 mμ range in (i) H_2SO_4-EtOH, 4:1, (ii) H_2SO_4-EtOH, 2:1 and (iii) conc. H_2SO_4 to identify and determine 1, 1-Δ^5-3β-hydroxysteroids. An analog computer has been used [42a] to study the effects of substituents on the ultraviolet spectra of 4-ene-3-one steroids.

α-Tocopherol was determined via ultraviolet spectra in ethanol after a preliminary separation [104a]. Absorbance ratios were used to test the efficacy of the separation. Nicotinamide was measured at 261 mμ and pyridoxine at 291 mμ by two-component spectrometry in 0.005 N HCl, and thiamine was determined at 246 mμ in 2 N HCl, all after ion-exchange chromatographic separation of the B-group vitamins on alginic acid [41]. The operational normality of α-chymotrypsin was determined by a photometric "titration" using the change in absorbance at either 310 or 335 mμ when enzyme acylated by N-*trans*-cinnamoylimidazole at pH 5 [124]. Vasilenko *et al.* [149a] used nine wavelengths in the 225–290 mμ region to determine the nucleotide composition of ribonucleic acids. Ellman [36a] used the biuret reaction for determination of proteins; changes in the ultraviolet spectrum near 260 mμ were used to determine peptide bonds. Creatinine in urine was determined at 234.5 mμ in pH 10.4 medium after an ion-exchange separation [1]. Wells and Wolken [153] used visible microspectrophotometry of *in situ* haemosiderin granules to show the similarity of the spectrum to that of protoporphyrin. Shichi and Hackett [126] presented ultraviolet and visible spectra of oxidized and $Na_2S_2O_4$-reduced cytochrome isolated from the mung bean (*Phaseolus aureus*).

Krisch [73] obtained visible spectra of microsomal cyto-
chromes of the hog and bovine adrenal gland. Ferredoxins as
electron carriers in photosynthesis were studied by Tagawa
and Arnon by 250–600 mμ spectra and by differential spectra
[140]. Characteristic ultraviolet (Soret) bands due to uro-
porphyrin and coproporphyrin were used by Zenker [161] for
the quantitative determination of the former and of total
porphyrins in urine without preliminary separation.

6.6 *Carbonyl compounds*: Harris and Zoch [54] determined fur-
fural in aqueous solution at 276 mμ; they noted the effect of
SO_2 and concluded that the system might also be useful
for the measurement of bisulfite. Mixtures of acetone and
diacetone alcohol (4-hydroxy-4-methylpentan-2-one) were
analyzed [10] in EtOH by absorbance measurements at 253
and 285 mμ, respectively.

6.7 *Cations*: Carnall [27] presented spectra in the range 350–2600
mμ of Tb^{3+}, Dy^{3+}, Ho^{3+}, Er^{3+}, Tm^{3+}, Yb^{3+}, Pr^{3+}, Nd^{3+}, Sm^{3+},
Eu^{3+}, all in molten $LiNO_3$–KNO_3 eutectic at 150°C. Nd^{3+} has
been determined differentially in nitric acid at 575 mμ [85].
Titanium was determined in ilmenite concentrates by differential spectra at 390 mμ [84]. Callahan [26] determined U^{6+}
from 4 to 40 ppm as the chloride complex in conc. HCl at
246 mμ. He found a coefficient of variation of less than 2%
at the 20 ppm level. Interference by 55 cations was cited.
Goldstein determined osmium by oxidation to Os^{8+}, extraction
into $CHCl_3$, and measurement at one of the following maxima
(ϵ is cited for each): 282 mμ, 1870; 289 mμ, 1760; 297 mμ,
1640; 304 mμ, 1400; 312 mμ, 1000. Only Cl^- and Ru^{8+}
interfered. The method applied to 0.4–3.0 mg Os with a
coefficient of variation of 3%. Alimarin and Gibalo [2a] measured niobium by a differential technique in HCl containing
tartrate at 281 mμ. Gold was determined by Murphy and
Affsprung at 323 mμ in the 7–40 ppm range, after treating
with tetraphenylarsonium chloride and extracting into $CHCl_3$
[106]. Martin and Bentley [86] studied visible spectra of V^{2+},
V^{3+}, and V^{4+} in various media. The ultraviolet and visible
spectra of $CoCl_2$ and $[CoCl_4]^{2-}$ in nitromethane, N,N-
dimethylformamide, dimethyl sulfoxide, and dimethylaceta-
mide were published [23]; interaction between solvent and

solute was observed. White [155], in Crouthamel's book, reviews spectrophotometric methods for uranium, thorium, zirconium, technetium, neptunium, plutonium, and americium.

6.8 *Detergents*: Hayashi [56] studied the effect of Tween 80 and of cetylpyridinium bromide upon the visible spectra of Congo Red at various pH values. Jungermann and Beck [66] used an ultraviolet technique to determine germicides in soaps and detergents.

6.9 *Drugs*: Kráčmar [72] reviewed the application of ultraviolet spectrometry to the evaluation of pharmaceutically active compounds; 270 references. Bush (25) noted that ultraviolet spectra at pH 10.5 were helpful in determining the nature and purity of barbiturates extracted from drugs and metabolic products. Tishler *et al.* [142] used a differential technique, pH 10 *vs.* pH 1.5, to estimate phenobarbital at 241 mμ. Stokes *et al.* [136] removed salicylate from urine by adsorption on Florisil, then used the differential absorption of 5,5-disubstituted malonylurea derivatives to estimate them (max. 252–255 mμ, min. 234–237 mμ, in 0.45 N NaOH; max. 238–240 mμ, no min. in pH 10–10.5 buffer). Vidic [150] heated morphine, codeine, and dihydrocodeine (all of which contain an alcoholic OH group) with 72% H_2SO_4 at 85°C for one hour to transform them into products having intense and distinctive ultraviolet spectra. Pernarowski *et al.* [111] used a clever absorbancy ratio technique to analyze the following binary and ternary mixtures; caffeine, phenacetin in $CHCl_3$, 250, 265.5 mμ; procaine HCl, amethocaine Cl in H_2O, 297.5, 311 mμ; benzocaine, procaine HCl in H_2O, 285, 311 mμ; phenylsalicylate, salicyclic acid, gentisic acid; quinine, brucine, strychnine. After chromatography on alumina, phenothiazine was measured [20] at 254.5 mμ in 10% by volume of petroleum ether–ethyl ether 3:1 in EtOH. Ultraviolet spectra were used [124a] to show the reduction of *p*-nitrobenzyl alcohol to α-hydroxy-6-sulfonium-*m*-toluene sulfonic acid.

Machek [82] determined salicylamide (I) in H_2O at 329 mμ; aspirin (II) in H_2O at 300 mμ; caffeine (III) in H_2O at 273 mμ by correcting for (I) and (II); and phenacetin in EtOH, by correcting for (I), (II), and (III).

Phenindione was measured [18a] at 288 mμ after extraction with toluene and dilution with ethanolic KOH.

Tablets, ointments, and creams were extracted [163] with diethylene glycol, diluted with pH 10 Britton and Robinson buffer, and measured at 373 mμ for nitrofurazone, and at 365 mμ for furazolidone; for nitrofurantoin, ethanol was substituted for diethylene glycol and the wavelength was 386 mμ.

Marzys [88] dealt with mixtures of sulfadiazine, sulfamerazine, and sulfathiazole.

Ultraviolet spectra of primycin, an antibiotic with strong antituberculotic effect, were published [139]; the same paper mentioned BEW (*n*-butanol-ethanol-water, 1 : 1 : 2) as a solvent for guanidine derivatives.

Katz [66a] determined streptomycin in feed at 324 mμ after acid extraction and conversion of the streptose moiety into maltol by a base.

Holbrook *et al.* [60] determined gibberellic acid in fermentation samples at 254 mμ after conversion to gibberellenic acid by holding for 75 min at 20°C in 27% aqueous HCl containing 10% EtOH. Isogibberellic, allogibberic, and gibberic acids do not interfere, but Sumiki's acid does.

Hexachlorophane and other phenols in pharmaceuticals were determined [37] by difference in absorbance between alkaline and acid or neutral solutions for 19 types of preparations. Dean *et al.* [32] used the pH dependence of the ultraviolet spectra of phenols and aromatic amines to determine disinfectants such as hexachlorophene, salicylanilide, *p*-phenylenediamine, and nitro-*p*-phenylenediamine. Jungermann and Beck [66] used ultraviolet spectra of eight different germicides in acidic and basic alcohol to estimate them after isolation from soaps and detergents by dimethylformamide extraction and evaporation. Halot [51] measured chloramphenicol at 275 mμ in ethanol after dissolving away a suppository excipient with light petroleum; he also determined naphazoline nitrate in medicaments at 281 mμ.

The now notorious thalidomide was determined [46] in body fluids by extraction and differential measurement at 220 mμ, N NaOH *vs.* 0.1 N HCl.

Machek [82] measured theophylline at 271 mμ and mersalyl acid at 291 mμ in dilute slightly acid solution. Theophylline was also determined [160] in tablets along with phenobarbitone by measurements at 252.5 and 274.5 mμ in pH 9.5 borate buffer.

The ultraviolet spectra of papaverine (I), papaverinol (II), and papaveraldine (III), and of their hydrochlorides in 96% EtOH and 0.1 M H$_2$SO$_4$ were presented [117a]. Those of (I) and (II) are similar, but mixtures of (I) and (III) can be analyzed at 335 and 285 mμ in 0.1 M H$_2$SO$_4$.

6.10 *Dyes*; Absorption spectra, 330–700 mμ, of more than 20 water-soluble dyes prohibited (in Spain) for use in foods were presented in acid, alkaline, and neutral media [150a].

Ultraviolet and visible spectra were helpful in the separation and identification of the colored and uncolored components of D&C Yellow No. 10 (sodium salt of sulfonated 2-(2-quinolyl)-1,3-indandione) [119].

Lachman *et al.* used reflectance spectra of tablets colored with FD&C Yellow No. 5, Blue No. 1, and Red No. 3, to evaluate the effect of the ultraviolet absorber, 2,4-dihydroxybenzophenone, upon their photostability [75].

Swartz *et al.* studied the influence of temperature and pH upon the surface color and the total dye content of tablets colored with FD&C Red No. 4, Blue No. 1, and Yellow No. 5. They studied the change of absorbance with time by the Guggenheim technique*. The first-order rate constant, K, for FD&C Yellow No. 5 at pH 5 and 60°C was found to be 5.6 × 10^{-3} days [137].

Visible spectra of Congo Red, Benzopurpurine 4B, Chrysamine G, and the 3,3'-disulfonic acid isomer of C.I. 22880 in dilute H$_2$SO$_4$ and acetic acid and (for the first three) at pH 2, 3, 5, and 7 were published [119a].

Hayashi [56] studied the effect of Tween 80 and cetylpyridinium bromide on the visible spectra of Congo Red at various pH values.

Levshin and Gorshkov [79] investigated the nature of the bonding forces of associated dye molecules in concentrated solution. In the case of Rhodamine C as concentration increases the absorbance at 555 mμ falls off, while the 525 mμ

---

*See E. A. Guggenheim, *Phil. Mag.* 2:538 (1926).

band grows; there is an isosbestic point at 535 mμ. Upon heating the short-wavelength absorbance diminishes, the long-wavelength band is partially restored. They concluded that the aqueous solutions are probably binary mixtures of monomers and associated molecules. Magdala Red behaved similarly. Crystal Violet solutions showed no isosbestic point and appeared to contain associated molecules of varying complexity.

The effect of solvent upon the aggregation of Methylene Blue, Rhodamine 6G, and pinacyanol iodide has also been studied via visible spectra [7]; aggregation was greatest in glycerol and ethylene glycol, less in acetone, acetonitrile, and formamide.

The sensitivity of the absorption spectra of merocyanines to solvents was discussed [104]; Coenen and Riester [29a] commented on absorption-(250–700 mμ)-structure relationships of branched polymethine dyes.

Three papers [110, 146, 147] discussed the photoreduction of Methylene Blue; two others [5] the reduction of Violanthrone B and Isoviolanthrone B.

A spectrometric method for the analysis of thioindigoids used isosbestic points in m-xylene solutions [110a].

Gerson and Heilbronner used spectra of p-dimethylamino-azobenzene and various derivatives to elucidate the structure of protonated forms [42b].

Burkhard [24] described the spectrometric estimation of the ionization constants of dyes.

Tables of extinctions of neutral, alkaline, and acid solutions of synthetic colors permitted as food dyes in Italy were published for the wavelength range 210–600 mμ [34].

6.11 *Fats and oils*: Franzke [42] presented ultraviolet spectra of some natural vegetable fats. Minutilli [95] reviewed the applications of spectrometry to the study of vegetable oils; 101 references. Schauenstein [123] described both possibilities and limitations of ultraviolet spectrometry in fat research. Both Morani [101] and Montefredine and LaPorta [99] used ultraviolet to characterize olive oils; the latter presented data on 206 samples.

Lauber [77] used ultraviolet measurements at 270 mμ of hexane solutions of cacao butter to detect extracted in expressed, when more than 10% of the former was present.

Autoxidized components of various wool waxes were measured by differential ultraviolet spectra [4].

6.12 *Foods*: Lee determined a caffeine in coffee and coffee mixtures by a three-wavelength method [77b]. Polzella [116] determined caffeine in roasted coffee; he reported $E_{1cm}^{1\%}$ in $CHCl_3 = 525$ at 274 mμ.

Ultraviolet was used to estimate antifermentatives (such as hydroxymethylfurfuraldehyde) in extracts of jam and wine [33].

Warfarin in aqueous solution was determined by Leahy and Waterhouse by ultraviolet spectrometry [77a]. The utility of ultraviolet in determining pesticide residues in foodstuffs was reviewed by Blinn [15].

Paar [110b] determined the drug Nitrofurazone (5-nitro-2-furaldehyde semicarbazone) in milk at 375 mμ after precipitating milk protein with sodium tungstate and H_2SO_4.

Zweig *et al.* [162] used ultraviolet measurements at 281 or 224 mμ as the last step in a long (extraction–esterification gas-liquid chromatography) procedure for sprout inhibitor (naphthalene acetic acid) residues in potatoes.

6.13 *N-Heterocyclics*: Nakanishi *et al.* published ultraviolet spectra of anions of uracils [107] and of 5,8-hydroxyisoquinolines and related compounds [108]. Snyder and Buell [131] determined basic N in gasoline to 0.01 ppm by ultraviolet absorption after separation by ion-exchange chromatography. They provided equations for total basic N (measured at 260 mμ) and pyridines to quinolines ratio (requiring data at both 260 and 300 mμ).

6.14 *Monocyclic aromatics*: Bouey and Yanari [18] discussed the effect of solvents upon the wavelength of maximum absorption of certain aromatics. Diyarov and Reizner [34a] used linear simultaneous equations to analyze a mixture of *o*-ethyltoluene (271 mμ), *m*-ethyltoluene (272 mμ) and *p*-ethyltoluene (273.5 mμ).

6.15 *Nitrogen compounds (other than amines, heterocyclics, and inorganics)*: Hahn and Jaffé [50] used spectra to study the rearrangement of 4-substituted azoxybenzenes to hydroxyazobenzenes in 100% H_2SO_4. Semba [125] presented absorption spectra of nitrobenzene derivatives. The effect of non-H-bonding solvents on the electronic spectra of aromatic N-oxides was studied [73a].

6.16 *Paper industry*: Pentosan was determined in woodpulp [65] by a differential ultraviolet technique which corrected for interference by hydroxymethylfurfuraldehyde.

Donetzhuber [34b] provided ultraviolet spectra of several paper pulps, glucose, mannose, xylose, arabinose, galactose, maltose, cellobiose, and *B*-methylglucoside dissolved in Cadoxen, $[Cd (en)_2] (OH)_2$.

Sato *et al.* [120b] determined residual lignosulfonic acid in unbleached sulfite pulp, estimating CH_3O at 280 mμ to obtain a lignin balance. Johnson *et al.* [62a] estimated lignin in wood at 280 mμ after a special purification of an acetyl bromide – acetic acid solution; they cited absorptivities for 18 species of wood. Ultraviolet was used to determine acid soluble lignin in sulfite pulp [79a] and to infer the presence of biphenyl groups in lignin [112].

6.17 *Phenols*: Baba presented electronic spectra of phenols [8] and naphthols [9]. Belova [14] determined monohydric phenols by ultraviolet absorption after chromatographic separation. Chen and Laidler determined pK values of phenol, cresols, and xylenols by ultraviolet spectrometry [28].

6.18 *Pigments, natural*: Absorption spectra of both normal and carotenoid-deficient cells of Chromatium, strain D, were reported [152]. Anderson and Calvin [3] found that the ratio of blue-band to red-band absorbance of very pure chlorophyll *a* made by a chromatographic technique designed to eliminate xanthophyll and pheophytin is 1.19. Wilson *et al.* [157] used tetracyanoethylene to remove the carotenoid absorption from solutions of pheophytin. Currie [31] used visible spectra of extracts of faeces of zooplankton to show the rapid breakdown of ingested chlorophyll *a*. Bruinsma [22] stated that in 80% acetone the spectra of α- and β-chlorophyll intersect at 651.8 mμ.

6.19 *Polycyclic aromatics*: Haenni *et al.* [49] devised a method for carcinogenic polynuclear hydrocarbons in mineral oil. Sensitivity is about 0.3 ppm. Reference absorbance is that of naphthalene at 275 mμ in "isooctane." The oil is first mixed with *n*-hexane, then extracted with dimethyl sulfoxide. The ultraviolet spectra of 10 of the 18 stereoisomers of *p,p'*-distyrylstilbene in tetrahydrofuran solution were compared with those of *p*-distyrylbenzene and *p*-styryltolan [96].

Monkman [98] reported the longest wavelength of maximum absorption of the monomethylbenzanthracenes. Sawicki *et al.* [121] published ultraviolet, visible, activation, and fluorescence spectra of a large number of polynuclear hydrocarbons.

The apparent ionization constants of fluorene and 119 other polycyclics were determined spectrometrically [45a].

Johnson [63] found 0.3–3% 5,6-benzidane at 321 mμ and about 4–5% dibenzofuran in commercial fluorene; he cited spectral data for pure fluorene.

Baba and Suzuki [9a] compared the electronic spectra of α- and β-anthrol with anthracene.

A paper on spectral response of photoconductivity in polycyclic aromatic compounds [120a] included visible spectra of thin films of violanthrene, perylene, anthanthrone, violanthrone, indanthrazine, and pyranthrene.

Keen *et al.* [71] reported an ultraviolet method for triphenylene, a radiolytic decomposition product of terphenyls, in polyphenyl reactor coolants.

6.20 *Polymers*: Lucchesi [80] reviewed the applications of spectroscopy in the protective coatings industry.

Luongo [81] presented ultraviolet spectra of Salol (phenyl salicylate) and of six antioxidants.

Antioxidants in plyethylene were detected by ultraviolet spectra of cyclohexane–ethanol solutions of the residue from an ether extract.

Stafford [133] determined Ionol (2,6-di-tert-butyl-*p*-cresol) in polyolefins by measuring at 365 mμ the colored oxidation product obtained by treating a cyclohexane extract with alkaline isopropyl alcohol. He found that interference from Santonox [4,4'-thiobis-(6-tert-butyl-*m*-cresol)] was not serious.

6.21 *Sulfur compounds*: Stahl [134] determined microgram amounts of divinylsulfone by measuring benzenethiol at 262 mμ before and after the addition of the sample (which precipitates *bis* (phenylthioethyl) sulfone).

Ultraviolet spectra were used as an aid to the deduction of the structure of dithiocarbamates [94].

6.22 *Water*: Goldman and Jacobs [44] determined NO_3^- in drinking water by acidifying with HCl and measuring at 220 mμ, correcting for organics by means of a reading at 272 mμ. Montgomery and Dymock [100] determined NO_3^- in fresh and saline waters by measuring a 4-cm thickness at 304 mμ after reacting sample with 2,6-xylenol in 75% H_2SO_4 in the presence of NH_4Cl. Sensitivity is 0.13 ppm nitrate nitrogen.

REFERENCES

1. Adams, W. S., Davis, F. W., and Hansen, L. E., *Anal. Chem.* 34:854 (1962).
2. Alderweireldt, F., *J. Chromatography* 5:98 (1961).
2a. Alimarin, I. P., Gibalo, I. M., and Ch'in, K.-J., *Zhur. Anal. Khim.* 17:60 (1962).
3. Anderson, A. F. H., and Calvin, M., *Nature* 194:285 (1962).
4. Anderson, C. A., and Wood, G. F., *Ibid.* 193:742 (1962).
5. Aoki, J., *Bull. Chem. Soc. Japan* 34:1817, 1820 (1961).
6. Arotsky, J., and Symons, M. C. R., *Nature* 193:678 (1962).
7. Arvan, K. L., and Zaitseva, N. E., *Optics and Specty.* 10 (2):137 (1961).
8. Baba, H., *Bull. Chem. Soc. Japan* 34:76 (1961).
9. Baba, H., and Suzuki, S., *Ibid.* 34:82 (1961).
9a. Idem., *Ibid.* 35:683 (1962).
10. Basiński, A., and Narebska, A., *Roczniki Chem.* 35:1131, 1381 (1961).
11. Bauman, R. P., Absorption Spectroscopy, John Wiley & Sons, New York, 1962.
12. Beaven, G. H., in: Advances in Spectroscopy, Vol. II, Thompson, H. W., (ed.), Interscience, New York, 1961.
13. Beaven, G. H., Johnson, E. A., Willis, H. A., and Miller, R. G. L., Molecular Spectroscopy, Macmillan Co., New York, 1962.
14. Belova, I. M., *Zuhr. Anal. Khim.* 16:229 (1961).
15. Blinn, R. C., in: Instrumental Methods for the Analysis of Food Additives, Butz, W. H. and Noebels, H. J. (eds.), Interscience, New York, 1961.
16. Blinn, R. C., and Gunther, F. A., *Analyst* 86:675 (1961).
17. Blumer, M., *Ibid.* 87:398 (1962).

18. Bouey, F. A., and Yanari, S. S., *Nature* 186:1042 (1960).

18a. Bose, B. C., and Vijayvargiya, R., *J. Pharm. and Pharmacol.* 14:58 (1962).

19. Breda, E. J., and Kotkas, E. V., *Anal. Chem.* 33:816 (1961).

20. Brierley, A., and Langbridge, D., *Analyst* 86:709 (1961).

21. Brode, W. R., and Corning, M. E., in: Physical Methods in Chemical Analysis. Vol. 1, second ed.. Berl. W. G. (ed.), Academic Press, New York, 1960.

22. Bruinsma, J., *Biochim. Biophys. Acta* 52:576 (1961).

23. Buffagni, S., and Dunn, T. M., *J. Chem. Soc.* p.5105, (1961).

23a. Bührer, R., and Reichstein, T., *Helv. Chim. Acta* 45:389 (1962).

24. Burkhard, R. K., *Trans. Kansas Acad. Sci.* 60:324 (1057); Chem. Abst. 56:995 (1962).

25. Bush, M. T., *Microchem. J.* 5:73 (1961).

26. Callahan, C. M., *Anal. Chem.* 33:1660 (1961).

27. Carnall, W. T., *Ibid.* 34:786 (1962).

28. Chen, D. T. Y., and Laidler, K. J., *Trans. Faraday Soc.* 58:480 (1962).

29. Cieleszky, V., and Nagy, F., *Z. Lebensm.-Untersuch. U.-Forsch.* 114:13 (1961).

29a. Coenen, M., Riester, O., *Ann. Chem.* 633, 110 (1960).

30. Crummett, W. B., *Anal. Chem.* 34:1147 (1962).

31. Currie, R. I., *Nature* 193:956 (1962).

32. Dean, D. E., Suffis, R., Levy, A., *Soap and Chem. Specialties* 37,(3):87, 89 101 (1961).

33. DeFrancesco, F., and Margheri, G., *Boll. lab. chim. provinciali (Bologna)* 12:5 (1961).

34. DeGori, R., and Grandi, F., *Ibid.* 12:60 (1961).

34a. Diyarov, I. N., and Reizner, M. S., *Zhur. Anal. Khim.* 17:102 (1962).

34b. Donetzhuber, A., *Svensk Papperstidn.* 64:898 (1961).

35. Dörr, F., and Held, M., *Angew. Chem.* 72(9):287 (1960).

36. Edward, J. T., and Wang, I. C., *Can. J. Chem.* 40:399 (1962).

36a. Ellman, G. L., *Anal. Biochem.* 3:40 (1962).

37. Elvidge, D. A., and Peutrell, B., *J. Pharm. and Pharmacol.* 13, Supp., 111T (1961).

38. Feldman, N., *Anal. Chem.* 34:256 (1962).

39. Feldmann, E. G., *J. Pharm. Sci.* 51(6):I(1962).

40. Fog, J., *Nature* 193:564 (1962).

41. Foster, J. S., and Murfin, J. W., *J. Pharm. and Pharmacol.* 13, Supp., 126T (1961).

42. Franzke, C., *Nahrung* 2:639 (1958).

42a. Garrett, E. R., Johnson, J. L., and Alway, C. D., *Anal. Chem.* 34:1472 (1962).

42b. Gerson, F., and Heilbronner, E., *Helv. Chim. Acta* 45:42, 51 (1962).

43. Glazer, A. N., and Smith, E. C., *J. Biol. Chem.* 236:2942 (1962).

44. Goldman, E., and Jacobs, R., *J. Am. Water Works Assoc.* 53:187 (1961).

45. Goldstein, G., U.S. Atomic Energy Comm., Rep. CF-59-6-43, 1959.

45a. Grantham, P. H., Weisburger, E. K., and Weisburger, J. H., *J. Org. Chem.* 26:1008 (1961).

46. Green, J. N., and Benson, B. C., *J. Pharm. and Pharmacol.* 13, Supp., 117T (1961).

47. Grinev, V. S., Rau, O. I., and Svischchev, G. M., *Optics and Spectry.* 11(4):263 (1961).

48. Grum, F., and Scharf, P. T., *J. Opt. Soc. Am.* 50:816 (1960).
49. Haenni, E. O., Joe, F. L., Jr., Howard, J. W., and Leibel, R. L., *J. Assoc. Offic. Agr. Chemists* 45:59 (1962).
50. Hahn, C. S., and Jaffé, H. H., *J. Am. Chem. Soc.* 84:946 (1962).
51. Halot, D., *Ann. pharm. franc.* 19:477, 483 (1961).
52. Hamilton, P. B., *Anal. Chem.* 34:3R (1962).
53. Hampel, B., Absorptionsspektroskopie im ultravioletten und sichtbaren Spektralbereich, Friedrich Vieweg & Sohn, Braunschweig, W. Germany, 1962.
54. Harris, J. F., and Zoch, L. L., *Anal. Chem.* 34:201 (1962).
55. Hartford, C. G., *Ibid.* 34:426 (1962).
56. Hayashi, M., *Bull. Chem. Soc. Japan* 34:119 (1961).
57. Hershenson, H. M., Ultraviolet and Visible Absorption Spectra Index for 1955–1959, Academic Press, New York, 1961.
58. Hirt, R. C., *Anal. Chem.* 34:276R (1962).
59. Hirt, R. C., Schmitt, R. G., Searle, N. D., and Sullivan, A. P., *J. Opt. Soc. Am.* 50:706 (1960).
60. Holbrook, A. A., Edge, W. J. W., and Baily, F., *Advances in Chem. Ser.* 28:159 (1961).
61. Holder, G. A., *Analyst* 86:679 (1961).
62. Jaffé, H. H., and Orchin, M., Theory and Application of Ultraviolet Spectroscopy, John Wiley & Sons, New York, 1962.
62a. Johnson, D. B., Moore, W. E., and Zank, L. C., *Tappi* 44:793 (1961).
63. Johnson, E. A., *J. Chem. Soc.* p. 994, (1962).
64. Johnson, G. D., *Spectrochim. Acta* 16:1489 (1960).
65. Jones, H. L., *Tappi* 44:745 (1961).
66. Jungermann, E., and Beck, E. C., *J. Am. Oil Chemists' Soc.* 38:513 (1961).
66a. Katz, S. E., *J. Ag. Food Chem.* 10:39 (1962).
67. Kaye, W. I., *Anal. Chem.* 34:287 (1962).
68. Kaye, W. I., *Analyzer* (Beckman Instruments, Inc.). 2(3):4 (1961).
69. Kaye, W. I., *Appl. Spectry.* 15:89, 130 (1961).
70. Kaye, W. I., and Poulson, R., *Nature* 193:675 (1962).
71. Keen, R. T., Baxter, R. A., Miller, L. G., Sheperd, R. C., and Rotheram, M. A., U.S. Atomic Energy Comm., Rep. NAA-SR-4356, 1961.
72. Kráčmar, J., *Pharmazie* 16:341 (1961).
73. Krisch, K., *Nature* 193:982 (1962).
73a. Kubota, T., and Yamakawa, M., *Bull. Chem. Soc. Japan* 35:555 (1962).
74. Labhart, H., *Helv. Chim. Acta* 44:447, 457 (1961).
75. Lachman, L., Urbanyi, T., Weinstein, S., Cooper, J., and Swartz, C. J., *J. Pharm. Sci.* 51:321 (1962).
76. Lang. L. (ed.), Absorption Spectra in the Ultraviolet and Visible Region, Vol. 1 (two parts), Academic Press, New York, 1961.
77. Lauber, E., *Mitt. Lebensm. u. Hyg.*, Bern 52:116 (1961).
77a. Leahy, J. S., and Waterhouse, C. E., *Analyst* 85:492 (1960).
77b. Lee, K.-T., *Ibid.* 86:825 (1961).
78. Lempicki, A., *J. Opt. Soc. Am.* 51:35 (1961).
79. Levshin, L. V., and Gorshkov, V. K., *Optics and Spectry.* 10:401 (1961).
79a. Lorås, V., and Löschbrandt, F., *Norsk Skogind.* 15:302 (1961).
80. Lucchesi, C. A., *Offic. Dig., Federation Paint & Varnish Production Clubs* 30:212 (1958).

81. Luongo, J. P., *Anal. Chem.* 33:1816 (1961).
82. Machek, G., *Sci. Pharm.* 29:73 (1961).
82a. Idem., *Ibid.* 29:257 (1961).
83. Malinin, D. R., and Yoe, J. H., *J. Chem. Educ.* 38:129 (1961).
84. Malyutina, T. M., and Dobkina, B. M., *Zavodskaya Lab.* 27:650 (1961).
85. Idem., *Ibid.* 27:653 (1961).
86. Martin, E. L., and Bentley, K. E., *Anal. Chem.* 34:354 (1962).
87. Maruyama, A., Shimoji, M., and Niwa, K., *Bull. Chem. Soc. Japan* 34:1243 (1961).
88. Marzys, A. E. O., *Analyst* 86:460 (1961).
89. Mason, S. F., *Quart. Revs.* 15:287 (1961).
89a. McAloren, J. T., *Nature* 195:797 (1962).
90. McDonald, F. R., and Cook, G. L., *Appl. Spectr.* 15:110 (1961).
91. McGrath, W. D., Pickering, W. F., Magee, R. J., and Wilson, C. L., *Talanta* 8:892 (1961); 9:227, 239 (1962).
92. Mellon, M. G., and Boltz, D. F., *Anal. Chem.* 34:232R (1962).
93. Milazzo, G., *Rend. ist. super sanità* 23:133 (1960).
94. Miller, D. M., and Latimer, R. A., *Can. J. Chem.* 40:246 (1962).
95. Minutilli, F., *Rass. Chim.* 10(3):24 (1958).
96. Misumi, S., Kuwana, M., and Nakagawa, M., *Bull. Chem. Soc. Japan* 35:135, 143 (1962).
97. Miyamoto, S., and Brochmann-Hanssen, E., *J. Pharm. Sci.* 51:552 (1962).
98. Monkman, J. L., *Appl. Spectry.* 16:22 (1962).
99. Montefredine, A., and LaPorta, L., *Olii Minerali Grassi Saponi Colori Vernici* 36:31, 63 (1959).
100. Montgomery, H. A. C., and Dymock, J. F., *Analyst* 87:374 (1962).
101. Morani, V., *Olii Minerali Grassi Saponi Colori Vernici* 37:327 (1959).
102. Morton, R. A., *J. Roy. Inst. Chem.* 84:5 (1960).
103. Morton, R. A., *Nature* 193:314 (1962).
104. Mostoslaoskii, M. A., and Izmail'skii, V. A., *Doklady Akad. Nauk SSSR* 142:600 (1962).
104a. Mulder, F. J., and Keuning, K. J., *Rec. trav. chim.* 80:1029 (1961).
105. Murnieka, R., and Gonter, C. E., *Anal. Chem.* 34:197 (1962).
106. Murphy, J. W., and Affsprung, H. E., *Ibid.* 33:1658 (1961).
107. Nakanishi, K., Suzuki, N., and Yamazaki, F., *Bull. Chem. Soc. Japan* 34:53 (1961).
108. Nakanishi, K., Ohashi, M., Kumasaki, S., and Koike, H., *Ibid.* 34:533 (1961).
109. Niebergall, P. J., and Mattocks, A. M., *Drug Standards* 28(3):61 (1960).
110. Obata, H., *Bull. Chem. Soc. Japan* 34:1057 (1961).
110a. Oksengendler, G. M., and Gerasimenko, Y. E., *Sbornik Statei, Nauch.-Issledovatel, Inst. Org. Poluprod. i Krasitelei* No. 2, 215 (1961).
110b. Paar, G. E., *J. Agr. Food Chem.* 10:291 (1962).
111. Pernarowski, M., Knevel, A. M., and Christian, J. E., *J. Pharm. Sci.* 50:943, 946 (1961); 51:688 (1962).
112. Pew, J. C., *Nature* 193:250 (1962).
113. Phillips, J. P., *Anal. Chem.* 34:171 (1962).
114. Pillion, E., Rogers, M. R., and Kaplan, A. M., *Ibid.* 33:1715 (1961).
115. Platt, J. R., *J. Chem. Phys.* 34:862 (1961).

116. Polzella, L., *Boll. lab. chim. provinciali (Bologna)* 12:23 (1961).
117. Price, W. C., in: Advances in Spectroscopy, Vol. I, Thompson, H. W. (ed.), Interscience, New York, 1959.
117a. Rácz, I., and Varsányi, D., *Pharm. Zentralhalle* 101:18 (1962).
118. Rao, C. N. R., Ultraviolet and Visible Spectroscopy, Butterworths, London, 1961.
119. Ritchie, C. D., Wenninger, J. A., and Jones, J. H., *J. Assoc. Offic. Agr. Chemists* 44:733 (1961).
119a. Roseira, A. N., and Tolmasquin, E., *Anais acad. brasil cienc.* 30:311 (1958).
120. Ross, S. D., and Wilson, D. W., *Analyst* 85:51 (1960).
120a. Sano, M., and Akamatu, H., *Bull. Chem. Soc. Japan* 35:587 (1962).
120b. Sato, K., Kobayashi, A., and Mikawa, H., *Ibid.* 35:662 (1962).
121. Sawicki, E.. Hauser, T. R., and Stanley, T. W., *Int. J. Air. Pollut.* 2:253 (1960).
122. Sawyer, R., *Instrum. & Control Systems* 34:2049 (1961).
123. Schauenstein, E., *Nahrung* 3:1123 (1959).
124. Schonbaum, G. R., Zerner, B., and Bender, M. L., *J. Biol. Chem.* 236:2930 (1961).
124a. Schroeter, L. C., and Higuchi, T., *J. Pharm. Sci.* 51:888 (1962).
125. Semba, K., *Bull. Chem. Soc. Japan* 33:1640 (1960); 34:722 (1961).
126. Shichi, H., and Hackett, D. P., *Nature* 193:776 (1962).
127. Shlyapochnikov, V. A., and Slovetskii, V. I., *Optics and Spectry.* 10:132 (1961).
128. Shmulyakovskii, Y. E., USSR Pat. 134907 (October 1, 1961).
129. Sill, C. W., *Anal. Chem.* 33:1584 (1961).
130. Slifkin, M. A.. *Nature* 195:693 (1962).
131. Snyder, L. R., and Buell, B. E., *Anal. Chem.* 34:689 (1962).
132. Srivastava, R. D., Van Buren, P. R., and Gesser, H., *Ibid.* 34:209 (1962).
133. Stafford, C., *Ibid.* 34:794 (1962).
134. Stahl, C. R., *Ibid.* 34:980 (1962).
135. Stanford, G. S., *J. Opt. Soc. Am.* 51:773 (1961).
136. Stokes, D. M., Camp, W. J. R., and Kirch, E. R., *J. Pharm. Sci.* 51:379 (1962).
137. Swartz, C. J., Lachman, L., Urbanyi, T., Weinstein, S., and Cooper, J., *J. Pharm. Sci.* 51:326 (1962).
138. Swinehart, D. F., *J. Chem. Educ.* 39:333 (1962).
139. Szilágyi, I., Vályi-Nagy, T., Szabó, I., and Keresztes, T., *Nature* 193:243 (1962).
140. Tagawa, K., and Arnon, D. I., *Nature* 195:537 (1962).
141. Tereshin, G. S., *Zhur. Anal. Khim.* 14:388, 516 (1959).
142. Tishler, F., Worrell, L. F., and Sinsheimer, J. E., *J. Pharm. Sci.* 51:645 (1962).
143. Tunnicliff, D. D., and Hawes, R. C., *J. Opt. Soc. Am.* 50:1039 (1960).
144. Ulrich, W. F., *Analyzer* (Beckman Instruments, Inc.) 3(2):3 (1962).
145. Underwood, A. L., and Howe, L. H., III, *Anal. Chem.* 34:692 (1962).
146. Usui, Y., and Koizumi, M., *Bull. Chem. Soc. Japan* 34:1651 (1961).
147. Usui, Y., Obata, H., and Koizumi, M., *Ibid.* 34:1049 (1961).
148. Vandenbelt, J. M., *J. Opt. Soc. Am.* 51:802 (1961).
149. Van Dranen, J., *Chem. Weekblad* 54:69 (1958).
149a. Vasilenko, S. K., Kamzolova, S. G., and Knorre, D. G., *Biokhimiya* 27:142 (1962).

150. Vidic, E., *Arzneimittel-Forsch.* 11:408 (1961).

150a. Villanua, L., Carballido, A., Olmedo, R. G., and Valdehita, M. T., *Anales bromatol. (Madrid)* 13:59 (1961).

151. Waggoner, W. H., and Chambers, M. E., *J. Org. Chem.* 26:2981 (1961).

152. Wassink, E. C., and Kronenberg, G. H. M., *Nature* 194:553 (1962).

153. Wells, C. L., and Wolken, J. J., *Ibid.* 193:978 (1962).

154. Wettermark, G., *Ibid.* 194:677 (1962).

155. White, J. C., in: Analytical Chemistry, Vol. 2, Crouthamel, C. E. (ed.), Pergamon Press, New York, 1961.

156. Wilson, H., *Anal. Biochem.* 1:402 (1960).

157. Wilson, J. R., Nutting, M.-D., and Bailey, G. F., *Anal. Chem.* 34:1331 (1962).

158. Wood, W. A., and Gilford, S. R., *Anal. Biochem.* 2:589 (1961).

159. Wybourne, B. G., *J. Opt. Soc. Am.* 50:84 (1960).

160. Yokoyama, F., and Pernarowski, M., *J. Pharm. Sci.* 50:953 (1962).

161. Zenker, N., *Anal. Biochem.* 2:89 (1961).

162. Zweig, G., Archer, T. E., and Raz, D., *J. Agr. Food Chem.* 10:199 (1962).

163. Żyżyński, W., *Acta Polon. Pharm.* 18:365 (1962).

Irrelevant Absorption in Quantitative Ultraviolet Spectrometry

Robert G. White

Tests and Inspection Dept.
National Aniline Division
Allied Chemical Corporation
Buffalo, New York

1. INTRODUCTION

1.1 Webster's defines *irrelevant* as: "Not relevant, not applicable or pertinent; extraneous." While I prefer the term "irrelevant absorption," the expressions "background" and "extraneous absorption" are often used and are equally correct for designating the absorption by unknown impurities that often obscures the interpretation of absorption spectra.

1.2 All UV spectroscopists have examined spectra of crude materials and noted the presence of absorption not due to the substance being assayed. What follows is a description of some of the ways that other workers have corrected for this irrelevant absorption to obtain a reasonably accurate assay.

1.3 We owe a great debt to two groups who pioneered in this field: the early infrared spectroscopists who devised the *baseline* technique [12, 28, 36] and the vitamin A assayers, notably Morton and Stubbs [21, 22, 22a, 22b] who contrived geometric corrections for linear irrelevant absorption.

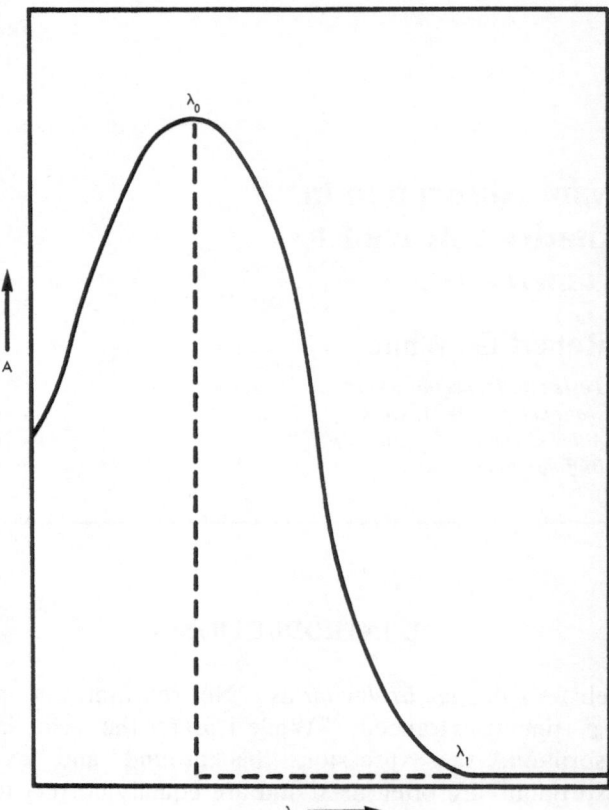

Fig. 1. Simple background correction.

2. SIMPLE BACKGROUND CORRECTION

2.1   If the sample spectrum is shifted downscale from zero absorbance at a wavelength λ_1 (usually longer than λ_0, the absorption maximum) where the pure material is virtually nonabsorbing, take $(A_{\lambda_0} - A_{\lambda_1})$ as a measure of concentration.

2.2   The A_{λ_1} may be due to turbidity, to "tars," or to other heterogeneous mixtures of impurities, or to even a shift in the spectrophotometer scale. Unless it is the latter, the extraneous absorbance is probably different (usually greater) at A_{λ_0} than at λ_1. However, the correction is easy to apply and usually better than none. (See Fig. 1.)

2.3   The American Society for Testing and Materials [3] recommends that this correction be made only when A_{λ_1} is equal to or less than $0.01A_{\lambda_0}$. This is a good operating rule, but need not be regarded as mandatory.

3.   SLOPE-TYPE BACKGROUND CORRECTION

3.1   This is a convenient means of compensating for irrelevant absorption, but it requires that the slope of the linear background be determinable and constant. Corrected A_{λ_0} is given by

$$A_{\lambda_0} \text{ (corr)} = A_{\lambda_0} - SA_{\lambda_1} \qquad (3.1)$$

where S = slope between λ_0 and λ_1 for the background, i.e., the ratio of background absorbance values at these λ's (see Fig. 2 [3].

4.   LINEAR BACKGROUND CORRECTION

4.1   This type of correction is treated more elaborately later in this report, but it is hard to better the ASTM presentation [3] for simplicity and clarity. Three wavelengths are used; λ_1 and λ_2 are on either side of λ_0, the absorption maximum.

4.2   In the special case where λ_1 and λ_2 are equidistant from λ_0, corrected A_{λ_0} is given by

$$A_{\lambda_0} \text{ (corr)} = A_{\lambda_0} - 0.5(A_{\lambda_1} + A_{\lambda_2}) \qquad (4.1)$$

4.3   The formula for the general case is

$$A_{\lambda_0} \text{ (corr)} = A_{\lambda_0} - \left[A_{\lambda_2} + (A_{\lambda_1} - A_{\lambda_2}) \left(\frac{\lambda_2 - \lambda_0}{\lambda_2 - \lambda_1} \right) \right] \qquad (4.2)$$

4.4   When Eq. (4.1) or (4.2) is used, the absorbance of the standard sample must be similarly corrected before absorptivity is calculated.

4.5   This technique is particularly useful when the absorption band is sharp and is flanked by nearby absorption minima. Obviously, the probability that the irrelevant absorption is approximately linear will increase with decreasing distance between λ_1 and λ_2.

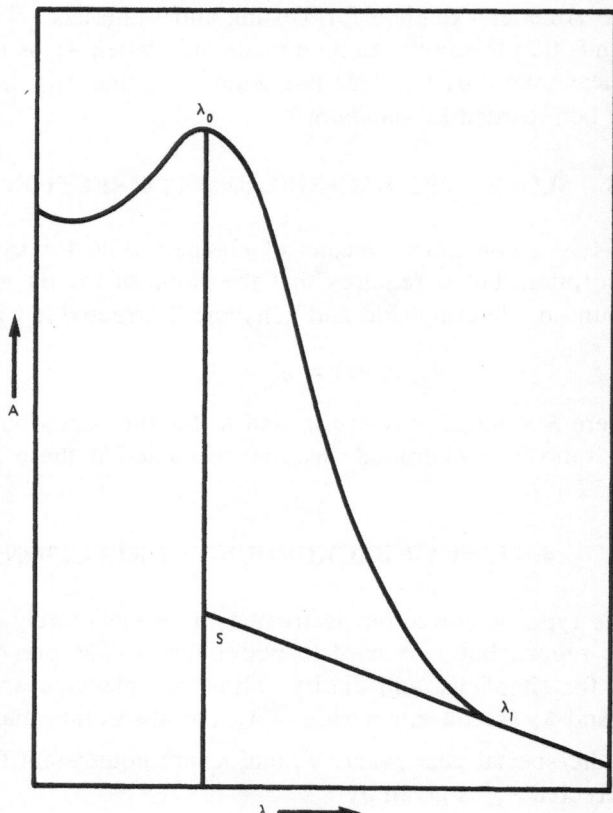

Fig. 2. Slope-type background correction.

4.6   This type of correction is difficult to apply when the slope
      from λ_2 to λ_1 is extreme or when the absorption maximum at
      λ_0 is broad and shallow.

5.   BASELINE METHOD

5.1   If the dotted line $\lambda_1 - \lambda_2$ is drawn on the spectrum and cor-
      rected A_{λ_0} is obtained by deducting the absorbance at the in-
      tersection (* in Fig. 3), we have the familiar "baseline" tech-
      nique [36].

5.2   Sometimes it is obvious on inspection that the irrelevant

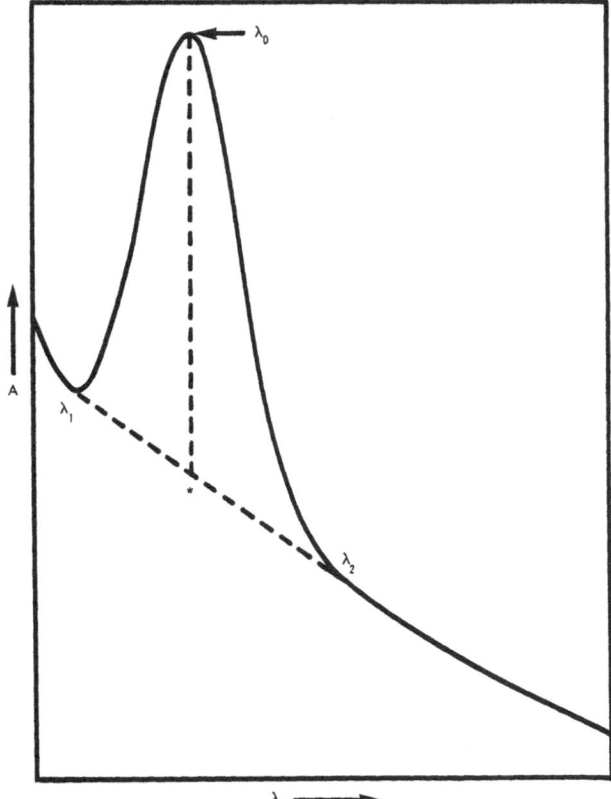

Fig. 3. Linear background correction.

absorption is curvilinear. If the analyst is audacious enough, he can construct a curved baseline which he believes will represent the "true" background.

5.3 The linear-baseline method is easy to use and teach and is probably the most widely used means of compensating for irrelevant absorption.

6. THE BANES–EBY METHOD

6.1 Banes and Eby [6] showed how to derive a linear background correction mathematically from Wright's baseline-type of correction [1]. Let λ_1 and λ_2 be minima, λ_0 a maximum (see Fig. 4); let x be the fraction of component sought; and let $1 - x$ be the fraction of background; then

$$A''_{\lambda_1} = xA_{\lambda_1} + (1 - x)A'_{\lambda_1} \qquad (6.1)$$

$$A''_{\lambda_0} = xA_{\lambda_0} + (1 - x)A'_{\lambda_0} \qquad (6.2)$$

$$A''_{\lambda_2} = xA_{\lambda_2} + (1 - x)A'_{\lambda_2} \qquad (6.3)$$

where A is the absorbance of the pure material, A' is the irrelevant absorption, and A'' is the sample absorbance.

Determine the values of A and A'' experimentally. Solve for x by

$$(1 - x)A'_{\lambda_0} = (1 - x)A'_{\lambda_2}$$
$$+ n\left[(1 - x)A'_{\lambda_1} - (1 - x)A'_{\lambda_2}\right] \qquad (6.4)$$

where

$$n = \frac{\lambda_2 - \lambda_0}{\lambda_2 - \lambda_1}$$

$$A'_{\lambda_0} = (1 - n)A'_{\lambda_2} + nA'_{\lambda_1} \qquad (6.5)$$

Substituting values of A'_{λ_1} and A'_{λ_2} from Eqs. (6.1) and (6.3) in Eq. (6.5), we obtain

$$A'_{\lambda_0} = \frac{\left[(1 - n)(A''_{\lambda_2} - xA_{\lambda_2}) + n(A''_{\lambda_1} - xA_{\lambda_1})\right]}{(1 - x)} \qquad (6.6)$$

Substituting A'_{λ_0} from Eq. (6.6) in Eq. (6.2) yields

$$x = \frac{A''_{\lambda_0} - (1 - n)A''_{\lambda_2} - n(A''_{\lambda_1})}{A_{\lambda_0} - (1 - n)A_{\lambda_2} - n(A_{\lambda_1})} \qquad (6.7)$$

6.2 The above, and much of what follows, assumes concentration c and thickness b to be held constant, so that we may deal with absorbances A without considering absorptivity a.

7.   THE MORTON–STUBBS CORRECTION.

7.1 The original Morton–Stubbs correction [22, 22a, 22b] for assay of vitamin A has been subjected to much criticism, discussion, revision, and simplification (for example in references [1, 4, 5, 8, 10, 11, 17–21, 23–26, 29, 34]).

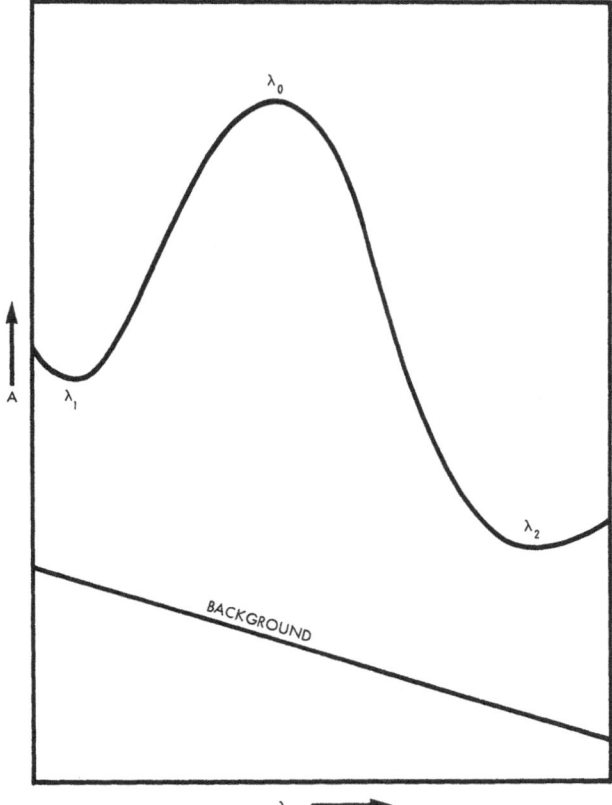

Fig. 4. Banes – Eby method.

It is a rather complicated two-stage correction: first one corrects for distortion in the shape of the true vitamin A curve; next, one corrects for vertical displacement due to nonvitamin A absorption.

The basic assumption for the first correction is that the irrelevant absorption is linear between two fixation points. These are the wavelengths on each side of the maximum at which absorbance is $6/7$ maximal.

7.2 Oser [23] published a simpler M—S correction to spectra in isopropyl alcohol.

7.3 Pancrazio and Duse [24] presented nomograms for both calculating and testing the M—S correction for vitamin A alcohol in isopropyl alcohol, and for vitamin A acetate in absolute ethanol.

This enables one both to apply the correction quickly and to ascertain whether the irrelevant absorption really is approximately linear between the two fixation points.

7.4 Prokhovnik [25] pointed out that the original M—S formula required the computation of three products to four significant figures and could not be handled on a slide rule. He proposed a different formula involving only two products.

7.5 Gridgeman [11] provided a comprehensive critique of geometric corrections for vitamin A absorption spectra, and gave a very clear derivation of the M—S correction. He described sources of error, presented hypothetical spectra to clarify the mechanism, showed the effect of using other fixation points than $^6/_7$ maximal, and discussed the validity of the method when applied to various vitamin-bearing oils.

7.6 Bagnall and Stock discussed the usefulness of the M—S correction [5]; they [4] and Adamson et al. [1] considered its reproducibility. Shaw and Jefferies [29] found the M—S correction to be disadvantageous for determination of ergosterol in yeast because ergosterol maxima are sharp and subsidiary λ's occur on steep slopes, inviting large absorbance errors due to small wavelength errors; they preferred a four-point correction using three maxima and a wavelength at which ergosterol itself had little absorption. Rogers [26] provided tables, Korr [17] a simpler formula, and Vigneau [34] a graphical procedure for using the M—S correction. Festenstein and Morton [10] applied a geometric correction in provitamin—D assay, while Lambertsen and Braekkan [18] used it in the determination of α—tocopherol. Lambertsen and Braekkan found that corrections not greater than $10-15\%$ were fairly reliable and that, in general, the correction procedure gave low results. Cama et al. [8] presented a very laborious 16λ procedure. Mariani and Gaudiano [19] showed a means of deriving a geometric correction.

7.7 The only vitamin A correction that we will consider in detail is that of McGillivray [20], who assumed the irrelevant absorp-

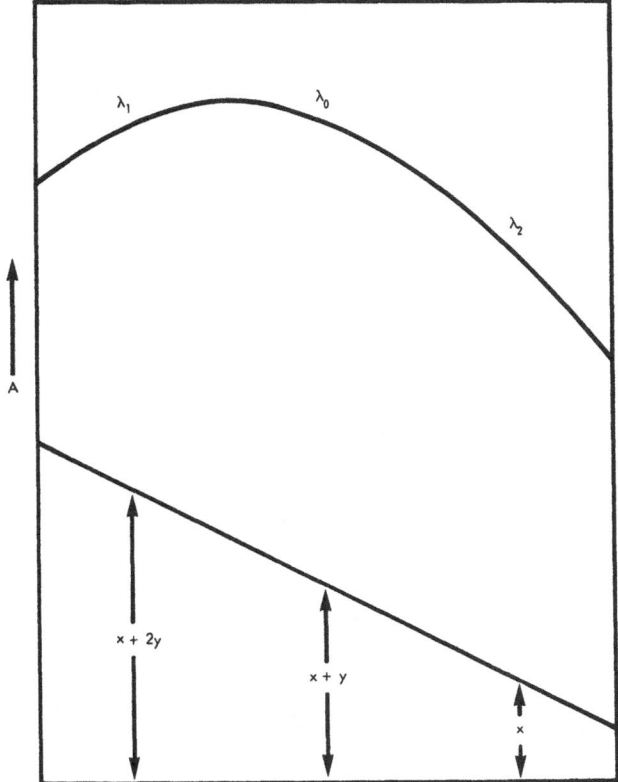

Fig. 5. McGillivray's method.

tion to be linear and used two fixation points λ_1 and λ_2 equidistant from the maximum* λ_0 (see Fig. 5). His derivation is interesting.

If A is the absorbance of pure material, and A'' is the sample absorbance, then

$$\frac{A_{\lambda_1}}{A_{\lambda_0}} = m \tag{7.1}$$

$$\frac{A_{\lambda_2}}{A_{\lambda_0}} = n \tag{7.2}$$

---

* Or from the analytical λ, whether or not it be a maximum.

$$\lambda_1 - \lambda_0 = \lambda_0 - \lambda_2 \tag{7.3}$$

$$A''_{\lambda_1} = (x + 2y) = m \left[A''_{\lambda_0} - (x + y) \right] \tag{7.4}$$

$$A''_{\lambda_2} - x = n \left[A''_{\lambda_0} - (x + y) \right] \tag{7.5}$$

Adding Eqs. (7.4) and (7.5)

$$A''_{\lambda_1} + A''_{\lambda_2} - 2(x + y) = m + n \left[A''_{\lambda_0} - (x + y) \right] \tag{7.6}$$

Solve in terms of $(x + y)$, the irrelevant absorption at λ_0, by

$$
\begin{aligned}
A''_{\lambda_1} + A''_{\lambda_2} &- 2(x + y) \\
&= (m + n)(A''_{\lambda_0}) - (m + n)(x + y) \tag{7.7}
\end{aligned}
$$

$$
\begin{aligned}
- \left[(m + n)(A''_{\lambda_0}) \right] &+ A''_{\lambda_1} + A''_{\lambda_2} \\
&= [2 - (m + n)](x + y) \tag{7.8}
\end{aligned}
$$

$$
\begin{aligned}
(x + y) = \frac{1}{2 - (m + n)} \\
\times \left\{ A''_{\lambda_1} + A''_{\lambda_2} - \left[(m + n)(A''_{\lambda_0}) \right] \right\} \tag{7.9}
\end{aligned}
$$

To use McGillivray's example: For pure vitamin A alcohol in EtOH, $\lambda_1 = 310$ mμ, $\lambda_0 = 325$ mμ, $\lambda_2 = 340$ mμ, $m = 0.846$, $n = 0.771$

$$(x + y) = 1/0.383 \left\{ A''_{\lambda_1} + A''_{\lambda_2} - \left[(1.617)(A''_{\lambda_0}) \right] \right\}$$

$$= 2.60 \left(A''_{\lambda_1} + A''_{\lambda_2} - 1.617 A''_{\lambda_0} \right)$$

7.8 A number of official methods for vitamin A assay employ an M—S-type correction, e.g., U.S.P. XVI [32] gets corrected absorbance (in isopropyl alcohol) by $6.815 A_{325\,m\mu} - 2.555 A_{310\,m\mu} - 4.260 A_{334\,m\mu}$.

8.   TUNNICLIFF'S EXTENSION OF BASELINE METHOD

8.1 Tunnicliff et al. [31] assumed that the irrelevant absorption in a system is not necessarily linear, but can be represented by some general analytic expression as a function of λ. They found useful both a power series and a sum of descending exponentials.

8.2 Calculations are formidable (or at least tedious), since they involve solving a number of simultaneous equations.

8.3 The number of terms used is critical. Too few may not be an adequate mathematical model. Too many may lead to the function's representing in part the spectra of the components to be determined.

8.4 A quartic proved suitable for interference encountered in the determination of benezene and toluene. However, a power series was not successful in the elimination of background from ethylbenzene and the xylenes, which have less complicated UV spectra than benzene. A sum of descending exponentials was used for these and for naphthalene.

9. THE SEVEN-POINT CORRECTION PROCEDURE

9.1 Ashton and Tootill [2] devised the seven-point method for the determination of the antibiotic griseofulvin in fermentation samples.

They found that the irrelevant absorption was never too complex to be described by a quadratic, while the absorption of the component sought was cubic or higher in nature.

Sets of wavelengths 2-mμ apart were picked, with the aid of orthogonal polynomials, to obtain the highest cubic component for pure griseofulvin. This turned out to be 288–300 mμ.

The procedure is thoroughly treated. Appendixes deal with loss of precision with increasing complexity of corrections and determination of the degree of the irrelevant absorption.

9.2 The Ashton and Tootill technique is so difficult that only the most determined worker would undertake to use it. Daly [9] has published a simpler version.

9.2.1 The absorption of a pure sample is described by the cubic

$$A_i = b_0 + b_1 x_i + b_{11} x_i^2 + b_{111} x_i^3 \tag{9.1}$$

The irrelevant absorption is given by the quadratic

$$A_i' = b_0' + b_1' x_i + b_{11}' x_i^2 \tag{9.2}$$

The absorption of a crude sample can be represented as

$$A_i'' = \rho A_i + A_i' \tag{9.3}$$

or by

$$A_i'' = (\rho b_0 + b_0') + x_i(\rho b_1 + b_1') \\ + x_i^2(\rho b_{11} + b_{11}') + x_i^3(\rho b_{111}) \tag{9.4}$$

where ρ is the proportion of pure substance in the crude samples.

If the A's are read for both the pure substance and the crude sample at six of seven equally spaced λ's, the middle one being omitted, one can construct Table I. Substituting the numerical values of x_i from the table in Eq. (9.4) yields

$$A_1'' = (\rho b_0 + b_0') - 3(\rho b_1 + b_1') \\ + 9(\rho b_{11} + b_{11}') - 27(\rho b_{111}) \tag{9.5}$$

$$A_2'' = (\rho b_0 + b_0') - 2(\rho b_1 + b_1') \\ + 4(\rho b_{11} + b_{11}') - 8(\rho b_{111}) \tag{9.6}$$

and so on, until

$$A_7'' = (\rho b_0 + b_0') + 3(\rho b_1 + b_1') \\ + 9(\rho b_{11} + b_{11}') + 27(\rho b_{111}) \tag{9.7}$$

TABLE I

λ x_i	Pure A_i	Crude A_i''	Irrelevant A_i'
-3	A_1	A_1''	A_1'
-2	A_2	A_2''	A_2'
-1	A_3	A_3''	A_3'
0	A_4	A_4''	A_4'
$+1$	A_5	A_5''	A_5'
$+2$	A_6	A_6''	A_6'
$+3$	A_7	A_7''	A_7'

Therefore

$$(A''_2 + A''_3 + A''_7) - (A''_1 + A''_5 + A''_6) = 36\rho b_{111} \qquad (9.8)$$

Doing the same for the A_i's yields

$$A_1 = b_0 - 3b_1 + 9b_{11} - 27b_{111}, \text{ etc.} \qquad (9.9)$$

whence

$$(A_2 + A_3 + A_7) - (A_1 + A_5 + A_6) = 36b_{111} \qquad (9.10)$$

To solve for ρ, one merely writes

$$\rho = \frac{(A''_2 + A''_3 + A''_7) - (A''_1 + A''_5 + A''_6)}{(A_2 + A_3 + A_7) - (A_1 + A_5 + A_6)} \qquad (9.11)$$

9.2.2 The middle-λ absorbance reading must be omitted. Daly says the best location for the omitted λ is halfway down the steep part of the slope (see Fig. 6). In the case of griseofulvin in ethyl acetate, the 7 λ's were 288, 290, 292, . . . , and 300 mμ.

10. CALDERBANK'S METHOD

10.1 The method of Calderbank, Morgan, and Yuen [7] is a linear background correction. While it is not particularly novel, the author's reasoning is easy to understand. Moreover, the original article contains an extremely clever geometric derivation.

10.2 Their problem was the determination of Diquat residues in potato tubers with 0.01 ppm sensitivity – Diquat is 1,1-ethylene-2,2'-bipyridylium dibromide dihydrate. It is the active ingredient in Reglone, a herbicicide and desiccant used in the United Kingdom to destroy potato haulm.

10.3 Spectra were taken of pure Diquat reduced with alkaline dithionite. Let λ_0 be 379 mμ, λ_1 be 375 mμ, and λ_2 be 385 mμ; then for pure reduced Diquat

$$\frac{A_{\lambda_0}}{A_{\lambda_1}} = 1.260$$

$$\frac{A_{\lambda_0}}{A_{\lambda_2}} = 1.535$$

$$\frac{A_{\lambda_1}}{A_{\lambda_0}} = 0.799$$

Therefore

$$\frac{A_{\lambda_0}'' - A_{\lambda_0}'}{A_{\lambda_1}'' - A_{\lambda_1}'} = 1.260 \tag{10.1}$$

$$\frac{A_{\lambda_0}'' - A_{\lambda_0}'}{A_{\lambda_2}'' - A_{\lambda_2}'} = 1.535 \tag{10.2}$$

Since the background is linear,

$$A_{\lambda_0}' - m\lambda_0 + c \tag{10.3}$$

$$A_{\lambda_1}' = m\lambda_1 + c \tag{10.4}$$

$$A_{\lambda_2}' = m\lambda_2 + c \tag{10.5}$$

where m and c are constants for the straight line. If c is eliminated,

$$A_{\lambda_1}' = A_{\lambda_0}' - m(\lambda_0 - \lambda_1) \tag{10.6}$$

$$A_{\lambda_2}' = A_{\lambda_0}' - m(\lambda_0 - \lambda_2) \tag{10.7}$$

Substituting these values for A_{λ_1}' and A_{λ_2}' in Eqs. (10.1) and (10.2) and solving for A_{λ_0}' yields

$$A_{\lambda_0}' = 2.28\, A_{\lambda_1}'' + 1.52\, A_{\lambda_2}'' - 2.79\, A_{\lambda_0}'' \tag{10.8}$$

Therefore, the corrected absorbance at λ_0 is

$$2.79\, A_{\lambda_0}'' - 2.28\, A_{\lambda_1}'' - 1.52\, A_{\lambda_2}'' \tag{10.9}$$

II. HISKEY'S CURVATURE–INVERSION TECHNIQUE

11.1 This is based upon the *variable reference* (see Fig. 7) technique, devised by Jones *et al.* [15] (who applied it to, among other things, the determination of FD&C Yellow No. 6 in the presence of the background absorption of caramel) and later used in vitamin assay by Schiaffino *et al.* [27].

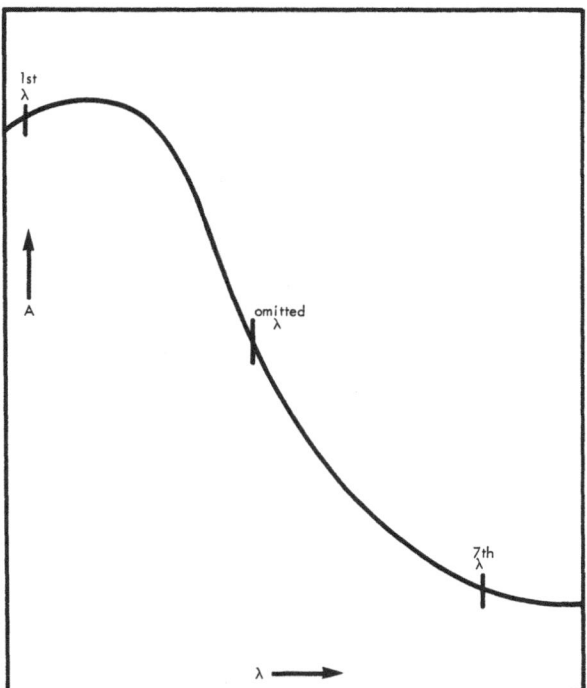

Fig. 6. Daly's method.

11.2 Basically, it consists of running a number of differential spectra, crude *vs.* purified, while varying the amount of the purified material in the reference beam of the spectrophotometer.

In the same manner as when the concentration of pure substance in the reference beam exceeds that in the sample beam, a negative absorption spectrum will be observed. The "curvature inversion" denotes that the end point of what amounts to a titration has been passed. If the increments of concentration change are small enough, the sample assay is quite accurate.

11.3 Hiskey applied this technique to a variety of pharmaceuticals. He argues that it is superior to other means of compensating for irrelevant absorption because it makes no assumptions about the spectral distribution of the background (which may vary from lot to lot).

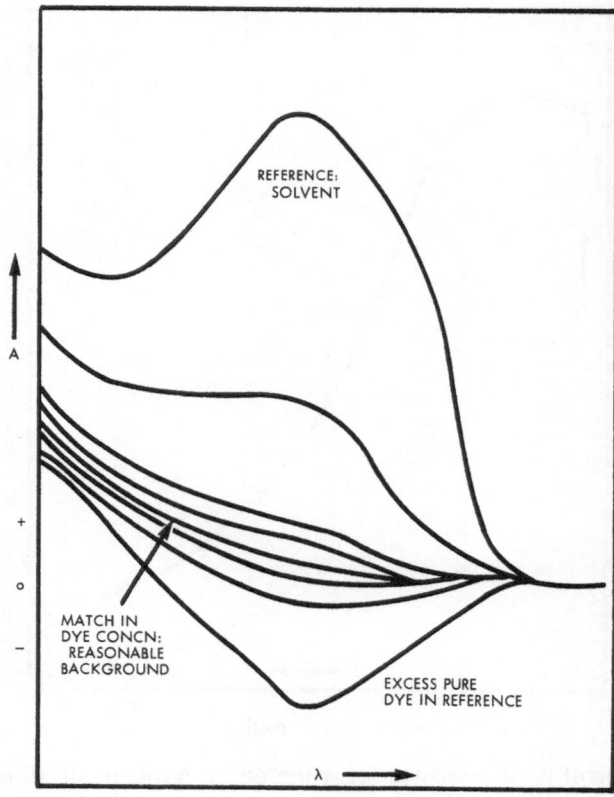

Fig. 7. Hiskey's curvature inversion.

11.4 Hiskey modified a Beckman DK-2 to compress the abscissa of the spectrum. This gave sharper curves and more precise estimates of equivalence point.

11.5 He proposes that an analog computer be used to simulate spectra of sample and pure substance by separate electronic function-generators. Difference curves could then be displayed oscillographically and the pure component function be varied *ad libitum* to generate the requisite difference spectra.

12. IMPURITY INDEX

12.1 Stearns' Impurity Index (II) [30] assumes a "horizontally absorbing impurity." Should the index λ be very far from the analytical λ, the assumption is risky. However, the *II* is a valuable tool for rating the relative quality of a series of samples contaminated in varying degrees by the same unknown impurities. Stearns points out that *II*'s are additive; this permits the blending of off-grade lots with better material to obtain a mixture of specified *II*.

12.2 As before, let the thickness b and concentration c be constant. Then

$$II = \frac{A''_{\lambda_1}}{A''_{\lambda_0}} - \frac{A_{\lambda_1}}{A_{\lambda_0}} \tag{12.1}$$

where A'' is the absorbance of a crude sample; A is the absorbance of a standard sample; λ_0 is the wavelength where concentration is measured; and λ_1 is the Impurity Index wavelength (λ_0 is usually a maximum; λ_1 is often a minimum; λ_1 is usually smaller than λ_0).

The *II* increases with diminishing purity. If *II* = 0, the quality of the sample equals the standard. Negative *II*'s are possible should the standard not be 100% pure.

12.3 To use the *II* in quantitative analysis, calculate α.

$$\alpha = \frac{II}{1 - (A_{\lambda_1}/A_{\lambda_0})} \tag{12.2}$$

$$\% = 100 \left[\frac{(1 - \alpha) A''_{\lambda_0}}{A_{\lambda_0}} \right] \tag{12.3}$$

12.4 If only an assay is required, and the *II per se* is not wanted, get percent by

$$\% = 100 \left[\frac{A''_{\lambda_0} - A''_{\lambda_1}}{A_{\lambda_0} - A_{\lambda_1}} \right] \tag{12.4}$$

12.5 Simpler techniques are in order when $A_{\lambda_1}/A_{\lambda_0}$ is very small:

$$\alpha \cong II \tag{12.5}$$

and represents, approximately, that fraction of the apparent concentration $A''_{\lambda_0}/A_{\lambda_0}$ which is actually due to the impurity.

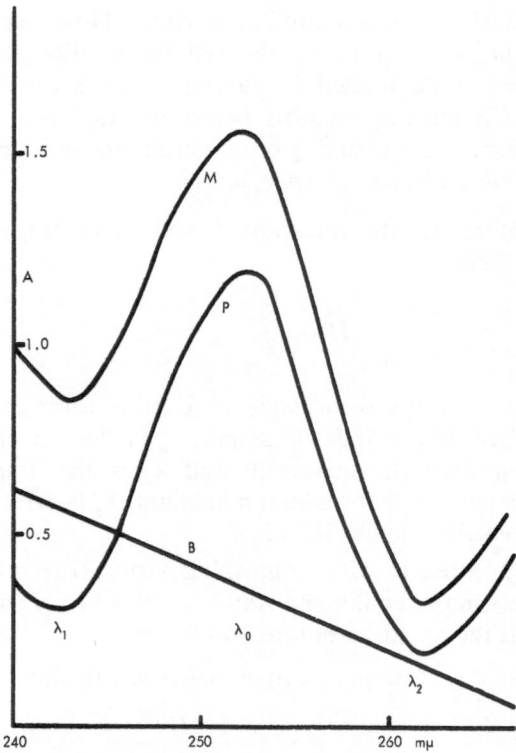

Fig. 8. Stearns' impurity index.

12.6 Figure 8 shows a hypothetical mixture spectrum M constructed by adding the linear background B to spectrum P.

The theoretical strength of M vs $P = 100\%$. The apparent strength read at λ_0 is $100\ (1.55/1.20) = 129\%$.

$$A_{\lambda_1} = 0.30 \qquad A_{\lambda_0} = 1.20$$
$$A''_{\lambda_1} = 0.85 \qquad A''_{\lambda_0} = 1.55$$

Applying Eqs. (12.1)–(12.3) yields

$$II = 0.298 \qquad \alpha = 0.397 \qquad \% = 78$$

(this result can also be obtained by applying Eq. (12.4) directly).

This result is 22% low i.e., we aren't much better off than we were with the incorrectly high result, 129%. Obviously, our irrelevant absorption is not "horizontal." The II can be used for quantitative work only when the background can be demonstrated to have essentially no slope from λ_0 to λ_1.

12.7 Stearns' Double Impurity Index (DII) [30] assumes the background to be linear, but not horizontal. The degree of slope doesn't matter, in theory, but curvature will cause errors.

12.8 For the DII select a third wavelength, λ_2, on the opposite side of λ_0 from λ_1. A long λ minimum or shoulder is a good choice for λ_2. Calculate α_d by

$$\alpha_d = \frac{w\,(II)_1 A_{\lambda_0} + (II)_2 A_{\lambda_0}}{w\,(A_{\lambda_0} - A_{\lambda_1}) + A_{\lambda_0} - A_{\lambda_2}} \tag{12.6}$$

$$w = \frac{\lambda_2 - \lambda_0}{\lambda_0 - \lambda_1}$$

12.9 In our example λ_0 is 252 mμ; λ_1 is 243 mμ; λ_2 is 263 mμ; A_{λ_0} is 1.20; A_{λ_1} is 0.30; A_{λ_2} is 0.20 and w is 1.22; $(II)_1$ we previously found to be 0.298; $(II)_2$, given that A''_{λ_2} is 0.335 and A''_{λ_0} is 1.55, is 0.049, therefore

$$\alpha_d = 0.236$$

From Eq. (12.3)

$$\% = 98.7$$

This is close to the 100% which would have been obtained had our mixture curve been more accurately drafted.

12.10 In many cases, a constructed baseline is just as effective and much easier. However, Stearns' and kindred methods are handy when we do not have the convenience of nearby absorption minima or shoulders.

12.11 Jennings and Edwards [14] used Stearns' DII method (they called it Mellon's) effectively in the UV determination of "methoxychlor," 2,2-*bis*-(*p*-methoxyphenyl)-1,1,1-trichloroethane on insecticide-treated pasteboard.

13. SPECTROPHOTOMETRIC PURITY INDEX [35]

13.1 The SPI (see Fig. 9) is akin to Stearns' II, but is not identical to it. One selects λ_0 and λ_1 in the same way and calculates

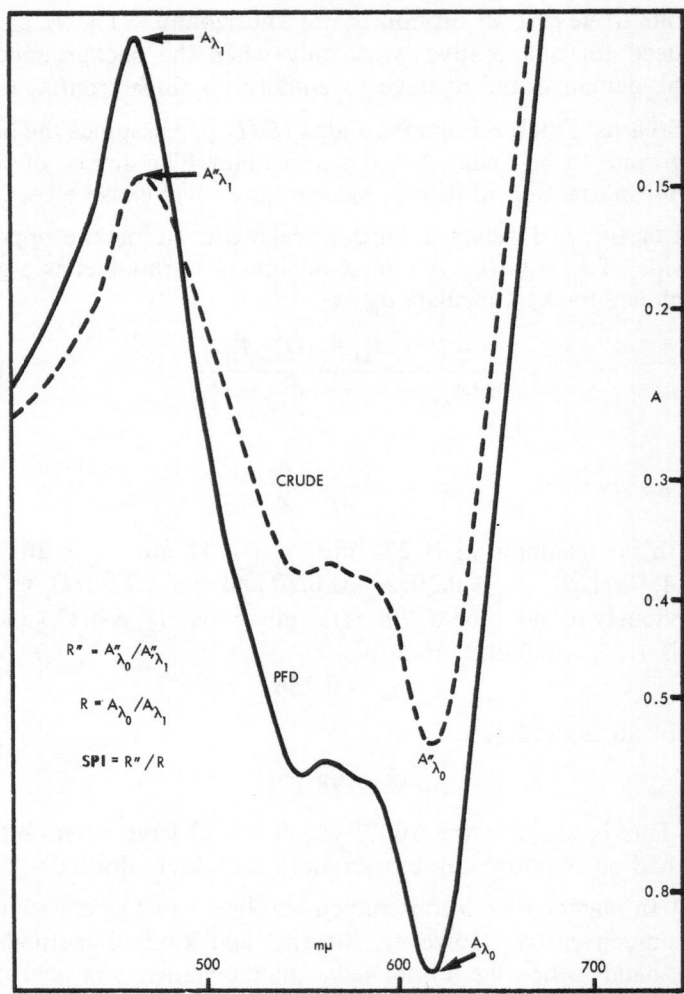

Fig. 9. Spectrophotometric purity index.

$$R'' = \frac{A''_{\lambda_0}}{A''_{\lambda_1}} \qquad (13.1)$$

$$R = \frac{A_{\lambda_0}}{A_{\lambda_1}} \qquad (13.2)$$

$$SPI = \frac{R''}{R} \qquad (13.3)$$

The *SPI* is not as versatile as Stearns' *II*, but has the property that it cannot exceed unity (if the purified sample is reliable).

13.2 The *SPI* can be used in quality specifications. It does not lend itself readily to quantitative estimation of irrelevant absorption, but can be used for ranking purity of successive productions of crude, or of fractions from countercurrent distribution separations, chromatographic columns, etc.

14. KIDDER'S COLOR INDEX

This is a good example of a do-it-yourself type of correction tailor-made for a particular task. Kidder [16] derived empirically a "color index," defined as $A_{440m\mu} - 0.8 A_{550m\mu} - 0.2 A_{360m\mu}$, to subtract background from the carotene-like coloring matter in raw rubber latex films and crepes.

15. THE CONSTRUCTION CORRECTION OF VANDENBELT AND SHEARER

15.1 This, the most recent paper [33] on irrelevant absorption that we have seen, may also prove the most interesting and useful. It assumes that turbidity and nonspecific absorption increase as λ decreases and that, over the range of interest, they do so exponentially.

15.2 In Fig. 10, the UV spectra of ethanolic solutions of product P, purified and crude, are given.

15.2.1 We select λ_0, a long-λ maximum; λ_1, an adjacent short-λ minimum (it need not *necessarily* be a minimum, nor λ_0 a maximum); λ_2, the shortest λ at which pure P is nonabsorbing.

15.2.2 Calculate $A_{\lambda_0}/A_{\lambda_1} = 1.54$

15.2.3 Calculate $A_{\lambda_0}'' - A_{\lambda_2}'' = 0.950 - 0.018 = 0.932$

15.2.4 Construct a line parallel to the abscissa passing through A_{λ_2}''

15.2.5 Erect perpendiculars at A_{λ_1}'' and A_{λ_0}''

15.2.6 Calculate: $(A_{\lambda_0}'' - A_{\lambda_2}'')/(A_{\lambda_0}/A_{\lambda_1}) = 0.932/1.54 = 0.605$

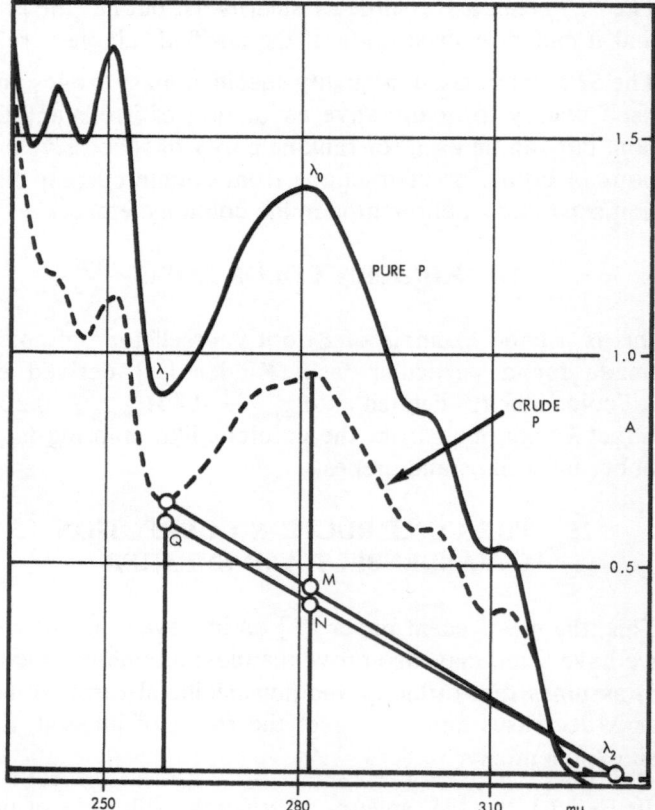

Fig. 10. Ultraviolet spectra of ethanolic solutions of product P.

15.2.7 To 0.605 add A''_{λ_2}; 0.605 + 0.018 = 0.623. Plot this point on the perpendicular from A''_{λ_1} and label it Q. Draw a line from Q to A''_{λ_2}. Label its intersection with the perpendicular from A''_{λ_0} as N.

15.2.8 Draw another line from A''_{λ_1} to A''_{λ_2}; label its intersection with the perpendicular from A''_{λ_0} as M.

15.2.9 Corrected A''_{λ_0} = observed $A''_{\lambda_0} - A''_{\lambda_2} - (M - N)$; in this case 0.950 − 0.018 − 0.031 = 0.901

15.2.10 In Fig. 10, theory assay of crude material = 65.0%. From uncorrected A''_{λ_0} we would get 68.3%. Correcting only for A''_{λ_2} the strength is 67.0%. Applying the Vandenbelt–Shearer correction, we obtain 64.8%.

16. CONCLUSION

Several different ways of compensating for irrelevant absorption have been described. Most of these assume the background to be linear. The same type of correction has been described more than once when different authors have given it their own particular treatments. When the same concept is approached from several routes at least one of them ought to be comprehended readily by every reader. The mathematics underlying the more elaborate techniques have not been dwelt upon: few present-day workers have time for lengthy calculations; those who do will want to consult the primary references.

REFERENCES

1. Adamson, D. C. M., Elvidge, W. F., Gridgeman, N. T., Hopkins, E. H., Stuckey. R. E., and Taylor, R. J., *Analyst* 76:445 (1951).
2. Ashton, G. C., and Tootill, J. P. R., *Ibid.* 81:225,232 (1956).
3. American Society for Testing and Materials, Rec. Practices for General Techniques of Ultraviolet Quantitative Analysis. E 169 – 63, 1964 Book of Standards, Part 31, 439.
4. Bagnall, H. H., and Stock, F. G., *Analyst* 77:356 (1952).
5. Bagnall, H. H., and Stock, F. G., *J. Pharm. and Pharmacol.* 4:81 (1952).
6. Banes, F. W., and Eby, L. T., *Anal. Chem.* 18:535 (1946).
7. Calderbank, A., Morgan, C. B., and Yuen, S. H., *Analyst* 86:569 (1961).
8. Cama, H. B., Collins, F. D., and Morton, R. A., *Biochem. J.* 50:48 (1951).
9. Daly, C., *Analyst* 86:129 (1961).
10. Festenstein, G. N., and Morton, R. A., *Biochem. J.* 60:22 (1955).
11. Gridgeman, N. T., *Analyst* 76:449 (1951).
12. Heigl, J. J., Bell, M. F., and White, J. U., *Anal. Chem.* 19:293 (1947).
13. Hiskey, C. F., *Anal. Chem.* 33:927 (1961).
14. Jennings, E. C., Jr., and Edwards, D. G., *Ibid.* 25:1179 (1953).
15. Jones, J. H., Clark, G. R., and Harrow, L. S., *J. Assoc. Offic. Agr. Chemists* 34:135 (1951).
16. Kidder, G. A., *Anal. Chem.* 26:311 (1954).
17. Korr, L., *Chemist Analyst* 42:15 (1953).
18. Lambertsen, G., and Braekkan, O. R., *Analyst* 84:706 (1959).
19. Mariani, A., and Gaudiano, A., *Rend. ist. super. sanità* 13:632 (1950).
20. McGillivray, W. A., *Anal. Chem.* 22:494 (1950).
21. Morton, R. A., *J. Pharm. and Pharmacol.* 2:129 (1950).
22. Morton, R. A., and Stubbs, A. L., *Analyst* 71:348 (1946).
22a. Morton, R. A., and Stubbs, A. L., *Biochem. J.* 41:525 (1947).
22b. Morton, R. A., and Stubbs, A. L., *Ibid.* 42:195 (1948).
23. Oser, B. L., *Anal. Chem.* 21:529 (1949).
24. Pancrazio, G., and Duse, V., *Analyst* 83:579 (1958).

25. Prokhovnik, S. J., *Ibid* 77:185 (1952).
26. Rogers, A. R., *Ibid* 80:903 (1955).
27. Schiaffino, S. S., Loys, H. W., Kline, O. L., and Harrow, L. S., *J. Assoc. Offic. Agr. Chemists* 39:180 (1956).
28. Seyfried, W. D., and Hastings, S. H., *Anal. Chem.* 19:298 (1947).
29. Shaw, W. H. C., Jefferies, J. P., *Analyst* 78:519 (1953).
30. Stearns, E. I., in:Analytical Absorption Spectroscopy, Mellon, M. G. (ed.), John Wiley & Sons, Inc., 1950, p. 390.
31. Tunnicliff, D. D., Rasmussen, R. S., and Morse, M. L., *Anal. Chem.* 21:895 (1949).
32. United States Pharmacopeia XVI, p. 938.
33. Vandenbelt, J. M., and Shearer, C. M., Carygraph 2 (1):8 (1962) (Applied Physics Corp., Monrovia, Calif.).
34. Vigneau, M., *Bull. soc. chim. biol.* 33:868 (1951).
35. White, R. G., in: Progress in Infrared Spectroscopy. Szymanski, H. (ed.), Plenum Press, 1962, p. 291.
36. Wright, N., *Ind. Eng. Chem., Anal. Ed.* 3:1 (1941).